NATIONAL GEOGRAPHIC

A

CONTENTS

Road Atlas
Adventure Edition

UNITED STATES · CANADA · MEXICO

Photo credits: (Cover) Hikers in Anza-Borrego State Park, California, Steve Casimiro; (This page) Bike touring on West Glacier Trail in Tongass National Forest, Alaska, Rich Reid/Colors of Nature.

B Contents and Legend

Road Map Table of Contents

U.S. States

Canadian Provinces and Territories

Road Map Legend

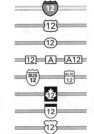

TRANSPORTATION

CONTROLLED ACCESS HIGHWAYS

Free
Toll; Toll Booth
Under Construction
Interchange and Exit Number
Ramp
Downtown maps only
Rest Area; Service Area
Ⓡ Ⓡ Ⓢ   Yellow with facilities

OTHER HIGHWAYS

Primary Highway
Secondary Highway
Multilane Divided Highway
Primary and secondary highways only
Other Paved Road
Unpaved Road
Check conditions locally

HIGHWAY MARKERS

Interstate Route
U.S. Route
State or Provincial Route
County or Other Route
Business Route
Trans-Canada Highway
Canadian Provincial Autoroute
Mexican Federal Route

OTHER SYMBOLS

Distances along Major Highways
Miles in U.S.; kilometers in Canada and Mexico
Tunnel; Pass
Driving Tour
Wayside Stop
One-way Street
Port of Entry
Airport
Official airport codes in parentheses
Auto Ferry; Passenger Ferry

RECREATION AND FEATURES OF INTEREST

National Park
National Forest; National Grassland
Other Large Park or Recreation Area
Small State or Provincial Park
with and without Camping
Public Campsite
Trail
Point of Interest
Visitor Information Center
Public Golf Course; Private Golf Course
Professional tournament location
Hospital
City maps only
Ski Area

CITIES AND TOWNS

National Capital; State or Provincial Capital
County Seat
State maps only
Cities, Towns, and Populated Places
Type size indicates relative importance
Urban Area
State and province maps only
Large Incorporated Cities

OTHER MAP FEATURES

JEFFERSON   County Boundary and Name
Latitude; Longitude
Time Zone Boundary
+ Mt. Olympus 7,965   Mountain Peak; Elevation
Feet in U.S.; meters in Canada and Mexico
Perennial; Intermittent River
Perennial; Intermittent or Dry Water Body
Dam
Swamp
Glacier

America's Best Adventure Destinations

A road map of America is, essentially, one big scenic route. No matter where you live, some segment of this country's easy-access natural wonders—desert buttes, raging white water, windswept tundra, tropical forests, and all the endless variations of mountains, seas, and plains this land has to offer—is just a drive away. It's not at all unusual for a San Franciscan to jump in her car and take off into the High Sierra for a weekend at Yosemite, while a Miamian might be strapping his kayak to the roof rack en route to the Everglades. Even Edward Abbey, the wilderness visionary, loved nothing more than to hit the road. He once wrote: "Whenever there was enough money for gas," to fill his pickup truck or cherry-red Cadillac, "we took off." • Where the pavement ends, the action begins. In that spirit, ADVENTURE's editors worked to assemble the 100 greatest hiking, biking, paddling, snow sports, climbing, and birding destinations in the United States in one ready-to-go volume. We've made it our business to bring you the best of the American outdoors. You'll find adventure itineraries not only for top-ten destinations like the Grand Canyon and the Smoky Mountains, but also for many of our lesser-known favorites, like Wisconsin's Kettle Moraine State Forest and Nevada's Ruby Mountains. Wherever you decide to go—whether it's a weekend jaunt an hour from home or a month-long, coast-to-coast drive—this atlas is your blueprint for action. Enjoy the journey.

The editors of

ADVENTURE
NATIONAL GEOGRAPHIC

present America's 100 top outdoor escapes

- Hiking and Backpacking
- Climbing
- Paddling
- Winter Sports
- Biking
- Birding

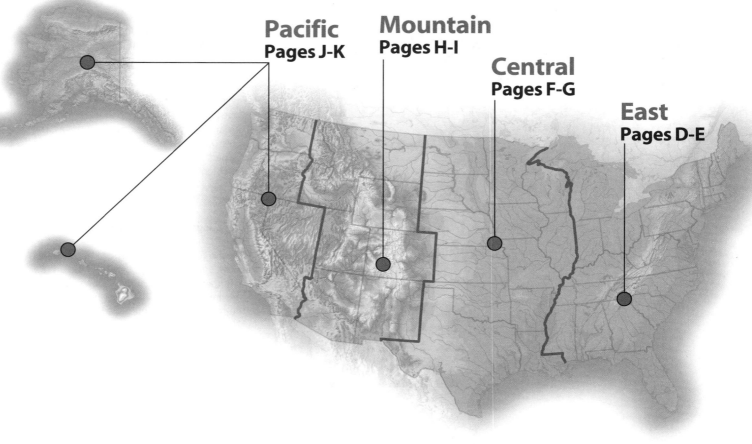

Pacific
Pages J-K

Mountain
Pages H-I

Central
Pages F-G

East
Pages D-E

EAST

The American wilderness ethic was born in the East. Here, surrounded by the subtle beauty of the Appalachians and the dunes and grassy coastal plains of the Eastern Seaboard, towering naturalists— Thoreau, Audubon, Emerson—pondered the role of nature in a new nation. The inspiration is still there. Today, the country's most extensively developed— and paved—tier offers travelers unparalleled access to adventure destinations of every kind, in a remarkable variety of protected lands. In places like Maine, Georgia, and North Carolina, wilderness beaches still abound. Mountain bikers devour the trail systems of West Virginia and Vermont. Climbers scale walls of granite, limestone, and sandstone from Maine to as far south as Alabama. Backpackers in 14 states can hike all or part of the nation's benchmark wilderness thruway: the Appalachian Trail. Here, a sampling of the best adventures, from day trips to epics, for the eastern road-tripper.

Adirondack Park

HIKING AND BACKPACKING
Towering lookouts to virgin trail

1 Cohos Trail, New Hampshire The Cohos Trail runs 170 miles from Crawford Notch to the Canadian border, and it may be the best, least-crowded trail in the entire region. A prime slice of the Cohos is the 25 miles between Jefferson and the South Pond Recreation Area. You'll top two 4,000-foot peaks, with 360-degree views into Maine and Canada. Visit www.cohostrail.org.

2 Delaware Water Gap National Recreation Area, Pennsylvania Hike an overlooked stretch of the Appalachian Trail between the Delaware Water Gap and Crater Lake (www.nps.gov/dewa), about two hours from Philadelphia. You'll pass Sunfish Pond, the Appalachian Trail's southernmost glacial lake. Overnight at the Mohican Outdoor Center, the southernmost camp operated by the Appalachian Mountain Club.

3 Great Smoky Mountains National Park, North Carolina and Tennessee. "Overwhelming" just begins to describe the 850 miles of hiking trails in Great Smoky Mountains National Park (www.nps.gov/grsm). To make the most of your visit, drive the Cades Cove Road, an 11-mile loop through a broad, verdant valley surrounded by mountains. The route offers some of the best opportunities in the park for wildlife viewing at a leisurely pace, with plenty of options for short hikes along the way. For visitor information, see the park's extensive website, or call 865.436.1200.

4 Talladega Mountain, Alabama Just an hour from both Birmingham and Atlanta, the Cheaha Wilderness in Talladega National Forest embraces stony, 2,407-foot Cheaha Mountain. For terrific summit views, day-hike the Pinhoti Trail to Cave Creek Trail on a ten-mile loop in Cheaha State Park (256.488.5111). Or weave a 17-mile, weekend-length circuit by combining sections of the Chinnabee Silent, Pinhoti, and Skyway Loop Trails (www.alapark.com/parks/cheaha-state-park).

CLIMBING
Trad climbing to razor-edge sport routes

5 Adirondack Park, New York The crowds follow the Van Hoevenberg Trail up Mount Marcy, the state's highest peak. Here's a better idea: Giant Mountain. The six-mile round-trip gains 3,000 feet, and every step brings more peaks into view. Or tackle a classic rock climb on Big Slide Mountain; just before the summit, veer off the trail and bushwhack to the base of Big Slide Cliff. The climbing is 5.7 to 5.9; the view, off the chart. Contact Adirondack Rock & River climbing guides (www.rockandriver.com, 518.576.2041).

6 Shawangunks, New York The bolt-banned mother lode of trad climbing—perhaps the country's best collection of easier routes—is one of the hotter bouldering areas as well. Trad climbers revel in the steep faces and gymnastic roofs of hundreds of named lines. High Exposure (5.6) was first climbed in the 1940s. Try Inverted Layback, a two-pitch 5.9 that often stops 5.10 leaders at the first-pitch crux. Rock and Snow, in New Paltz (845.255.1311), has more information (www.rockandsnow.com).

7 New River Gorge National Park, West Virginia Though developed originally by trad climbers, most come to the "New" these days for the huge cache of one-pitch bolted routes on dense sandstone, most rated 5.10 to 5.12. Legacy (5.11a) is the quintessential local climb. Insiders make their way to Mutiny (5.11c), a steep, razor-sharp arête at the edge of Summersville Lake (www.nps.gov/neri). Contact Water Stone Outdoors (www.waterstoneoutdoors.com) for more information (304.574.2425).

8 Red River Gorge, Kentucky With nearly a thousand routes in an area otherwise devoid of sheer rock faces, the "Red" is a standby for Southern and Midwestern climbers. The steep, heavily pocketed limestone walls are home to some of the stiffest routes in the East and are regularly visited by climbing greats from around the globe. Guides available through Red River Outdoors (www.redriveroutdoors.com, 606-663-7625).

Delaware Water Gap

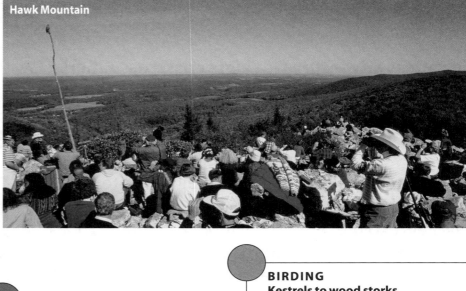
Hawk Mountain

PADDLING
Georgia white water to Everglades epic

9 **LaMoille River, Vermont** No northern Vermont road trip is complete without a day on the Lamoille River, near Stowe. The smooth water runs in the shadow of Mount Mansfield and underneath the 134-year-old Poland covered bridge. Umiak Outdoor Outfitters, in town, rents boats (www.umiak.com, 802.253.2317).

Youghiogheny River

10 **Youghiogheny River Gorge, Pennsylvania** This 1,700-foot-deep gash in the Laurel Highlands, about one and a half hours from Pittsburgh, is home to premier Eastern white water. Canoe the gentle 9-mile Middle Yough or tackle the Lower's Class III and IV rapids on a trip from White Water Adventurers (www.wwaraft.com, 724.329.8850).

11 **Chattooga River, Georgia** April is the height of Georgia's whitewater season. For some of the best rapids, head to the town of Clayton to join Southeastern Expeditions' rollicking seven-hour trip on the Chattooga (www.southeasternexpeditions.com, 800.868.7238), the river made famous by the film *Deliverance*.

12 **Gulf Islands National Seashore, Mississippi** Two hours east of New Orleans, near Ocean Springs, Mississippi, the Gulf Islands National Seashore is made up of 160 protected miles of sand, surf, and deserted isles (www.nps.gov/guis, 228.230.4100). A multiday canoe tour yields maritime forests filled with herons and gators, as well as undisturbed barrier islands that are home to more than 320 species of birds.

13 **Everglades National Park, Florida** Threading the vast and mysterious mangrove wetlands of the western Everglades for 99 miles, the Wilderness Waterway is one of the premier flatwater paddling routes in the United States (www.nps.gov/ever, 305.242.7700). A fortunate few kayakers or canoeists will glimpse bobcats, sea turtles, or manatees; all will be dazzled by the Everglades' bountiful bird life. Paddlers follow numbered markers and can stay overnight at beach campsites and atop raised platforms called chickees. ⬙

WINTER SPORTS
Ski-in dining to White Mountains ice

14 **Sugarloaf, Maine** For lift-accessed, above-treeline skiing, a rarity in the East, head to Maine's Sugarloaf (www.sugarloaf.com), where skiers can descend through broad snowfields. Come spring, the base lodge's outdoor sitting area, the Beach, is given over to sun worship. The Sugarloaf Inn offers the most affordable slopeside lodging (800.843.5623).

15 **White Mountains, New Hampshire** Two things rocky New England has in abundance are flowing water and cold temperatures. Combine them and you get nearly unlimited ice-climbing routes, including easily accessible classics scattered throughout New Hampshire's White Mountains. Learn the basics in a one- or two-day course conducted by Mooney Mountain Guides (www.mooneymountainguides.com, 603.545.2600). ⬙

16 **Whiteface Mountain, New York** It's not often you get to ski an Olympic mountain. At Lake Placid's Whiteface, the site of the 1980 Winter Games, 90 trails worm their way over the greatest vertical drop in the east—3,430 feet (www.whiteface.com). Try Excelsior, a perfect groomer for a morning warm-up, and then hit Skyward to test your mettle on a portion of the Olympic downhill. ⬙

17 **Porcupine Mountains Wilderness State Park, Michigan** Ski under the stars at Porcupine Mountains Wilderness State Park (www.michigan.gov/dnr, 906.885.5275), on Michigan's Upper Peninsula, as kerosene lanterns illuminate the one-mile Superior Loop Trail on winter Saturdays. Start the day with a glide across the 23-mile network of groomed cross-country trails, with a stop to dine at a ski-in log cabin within earshot of big winter waves slamming into Lake Superior's south shore.

BIKING
Rolling singletrack to deep-woods rock-hopping

18 **Killington, Vermont** With 30 miles of gondola-served trails, Killington is the largest mountain biking resort in the state (www.killington.com, 800.734.9435). Reward yourself after a long day of pedaling with a (motorized) sunset spin over the switchbacks of Lincoln Gap and Appalachian Gap roads to the village of Waitsfield, in the heart of the Mad River Valley. Check in at the charming Tucker Hill Inn, a 1940s-style ski lodge (www.tuckerhill.com, 802.496.3983). ⬙

19 **Elk River, West Virginia** When mountain bikers dream, it's of places like Elk River. Elk River Inn & Cabins (elkriverwv.com, 304.572.3771), located in Slatyfork, has a comfortable lodge, an in-house bike shop, and more than 200 miles of trails nearby—not to mention a gourmet restaurant. The 13-mile Bear Pen Trail loop includes some of the most scenic and varied singletrack in the 920,000-acre Monongahela National Forest.

20 **Asheville, North Carolina** Asheville is fast becoming the Boulder of the Southeast—with better riding. Pisgah National Forest, just beyond the city limits, boasts scores of trails. Tsali Recreation Area offers 41.8 miles of hard-packed singletrack and a minimum of bumps. Check out the Laurel Mountain loop for classic East Coast forest riding. The downhill portion is grade-A. Liberty Bicycles (libertybikes.com, 828.274.2453) has more info. ⬙

Asheville

21 **Piedmont Region, South Carolina/Georgia** They're little known, but the biking trails in the rolling, creek-threaded Piedmont north of Augusta, Georgia, rival any in the Southeast. The 24-mile out-and-back on the Wine-Turkey Creek Trail, northeast of Modoc, South Carolina, is a must-do. Check the Sumter National Forest Web site for details (www.fs.usda.gov/scnfs, 803.561.4000).

BIRDING
Kestrels to wood storks

22 **Hawk Mountain, Pennsylvania** Located on Pennsylvania's Kittatinny Ridge, Hawk Mountain Sanctuary (www.hawkmountain.org, 610.756.6961) is the oldest continuously operating hawk watch in the world. For good reason: 18,000 migrating hawks soar by each fall. Among the 16 species of raptors that are commonly seen are the eagle, kestrel, sharp-shinned and red-tailed hawks, black vulture, and golden eagle.

23 **Delaware Bay, Delaware/New Jersey** As one of four major estuaries in North America that serve as critical shorebird stopovers, Delaware Bay is rightfully a popular place. Visit May through June, when hundreds of thousands of migrating shorebirds—red knots, sanderlings, ruddy turnstones, and semi- palmated sandpipers included—come to feast on the eggs of spawning horseshoe crabs. Contact the Partnership for the Delaware Estuary (www.delawareestuary.org, 800.445.4935) for more information. ⬙

24 **Piedmont National Wildlife Refuge, Georgia** This 35,000-acre refuge, most of it loblolly pine forest, is a haven for nesting red-cockaded woodpeckers plus several species of warblers, nuthatches, and sparrows. May to early July is the high season, when the early morning calls are most impressive. Contact (www.fws.gov/refuge/Piedmont, 478.986.5441) for info.

25 **Big Cypress National Preserve, Florida** The cypress stands, marshlands, and mangrove forests of Big Cypress, the country's first national preserve, are a haven for wood storks, red-cockaded woodpeckers, short-tailed hawks, and several species of wading birds—if you can find them through the thick subtropical vegetation. To hedge your bets for a good sighting, visit in fall or spring, the most active seasons for foraging. The visitors center has weather information (www.nps.gov/bicy, 239.659.4111).

⬙ National Geographic topographic map available for this area.
800-962-1643
www.natgeomaps.com

CENTRAL

Countless adventure travelers crisscross America's midsection on their way to the dizzying topography of the West or the backwoods glories of the East. Let them hurtle by. Savvy climbers, paddlers, hikers, and—yes—skiers know the truth: The central region harbors a surprising variety of terrain, climate zones, and adventure potential. The Great Lakes, where kayakers explore remote, cave-pocked islands, boast more coastline than California. The area around Minnesota's Boundary Waters is one of the country's great all-season multisport playgrounds. The sprawling grasslands of the Dakotas inspired Theodore Roosevelt to create the world's first national park system—traveling through them you'll understand why. Across the Upper Midwest, extensive networks of well-maintained cross-country ski trails that lace thick woodlands are served by cozy lodges. Down south, snow quickly becomes a memory in places like Texas' mountain-studded Big Bend National Park or the wetlands of the Big Thicket National Preserve. Here, some signature adventures in America's heartland.

Wichita Mountains

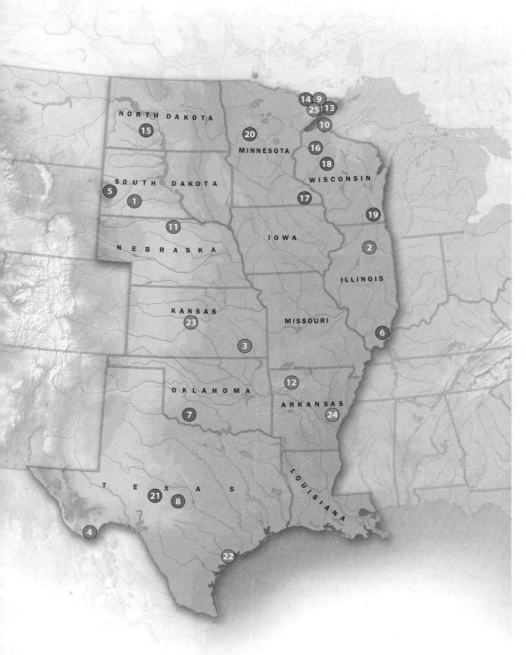

HIKING AND BACKPACKING
Borderlands to Badlands

1 **Badlands National Park, South Dakota** The 64,000-acre Badlands Wilderness is the largest trailless expanse in Badlands National Park (www.nps.gov/badl, 605.433.5361). Lose yourself (not literally, please) for a night or a week among the tortuous chains of eroded shale hills that gave Badlands its name, or enjoy on of the largest expanses of mixed-grass prairie—with more than 400 different species—in the U.S. ⬦

2 **Starved Rock State Park, Illinois** Just two hours from Chicago, 18 sandstone canyons and endless oak- and cedar-lined bluffs put a wrinkle in the Big Flat. Hike the riverside past the namesake Starved Rock, a 125-foot sandstone tower, where a group of cornered Illiniwek Indians met their untimely demise. Contact the visitors center for info (www.starvedrocklodge.com, 800.868.7625).

3 **Elk City Lake, Kansas** Even though it doesn't look like it, you'll still be in Kansas when you tackle the Elk River Hiking Trail. The route stretches 15 rocky miles along scenic Elk City Lake, outside of Wichita. Start at the west end of Elk City Dam and creep through the surrounding grasslands and hardwood forests to your camp atop a trailside bluff (ksoutdoors.com/State-Parks/Locations/Elk-City, 620.331.6295).

4 **Big Bend National Park, Texas** Local outfitter Desert Sports offers multisport excursions in the wildly diverse terrain of Big Bend (www.desertsportstx.com, 432.371.2727). A three-day biking and boating tour covers some of the most remote areas of the park, cycling through the historic Glenn Springs district before paddling the bottom of the "bend" in Big Bend. You'll experience nights as dark as coal, and see many of the hundreds of bird species that take refuge in the park. (www.nps.gov/bibe, 432.477.2251). ⬦

CLIMBING
The Needles to Enchanted Rock

5 **The Needles, South Dakota** Some of the best rock routes in the country are found in the Great Plains. At the Needles, in the Black Hills, hundreds of majestic granite spires are a climbers' playground. The action is centered at Custer State Park where you'll find anything from slab to crack climbs (gfp.sd.gov/parks/detail/custer-state-park, 605.255.4515). Go it with a partner, or contact the Sylvan Rocks climbing school for a guide (www.sylvanrocks.com, 605.484.7585). ⬦

6 **Shawnee National Forest, Illinois** Pretty but primitive Pharaoh Campground in southern Illinois's Shawnee National Forest (www.fs.usda.gov/shawnee, 618.253.7114) makes an ideal base camp for traipsing over, under, and around the bulbous sandstone formations of the Garden of the Gods, an area that would seem at home in Utah's tricked-out rockscapes. Local crag Jackson Falls features hundreds of climbing routes for those who want to go the extra inch.

7 **Wichita Mountains Wildlife Refuge, Oklahoma** Hundreds of climbing routes rise from the plains of Oklahoma's Wichita Mountains Wildlife Refuge (www.fws.gov/refuge/wichita_mountains, 580.429.2197). To find your way around, contact Guide for a Day and order *Oklahoma Rock: A Climber's Guide* $35; (www.guideforaday.com). That night, sleep at Doris Campground, then drive to Meers Store and Restaurant to refuel with peach cobbler (facebook.com/MeersBurger, 580.429.8051).

8 **Enchanted Rock State Natural Area, Texas** Squint real hard and Enchanted Rock in Texas Hill Country resembles Yosemite's Lembert Dome. Fingers won't detect the difference—this pink granite dome encourages amazing feats of friction. Among the many top-rope and multi-pitch routes, you'll find crack systems alongside low- to high-angle slabs. For info, contact Enchanted Rock State Natural Area (tpwd.texas.gov/state-parks/enchanted-rock, 830.685.3636).

Big Bend

PADDLING
Southern hollers to northern lights

9 Boundary Waters Canoe Area Wilderness, Minnesota Permits for the Boundary Waters Canoe Area Wilderness (www.fs.usda.gov/superior, 218.626.4300), so hard to find during summer high season, go begging during much of the shoulder seasons. Launch at Lake One and paddle to lovely Clearwater Lake, where walleye bite and moose rut. With an extra day, complete a 28-mile circuit through Turtle, Bald Eagle, and Gabbro Lakes, returning via South Kawishiwi River. ◈

10 Apostle Islands National Lakeshore, Wisconsin Tucked in the southwestern corner of Lake Superior, the 21 Apostle Islands (www.nps.gov/apis, 715.779.3397) feature some of the most remarkable sea kayaking in the Midwest. Stockton Island, the most popular in the archipelago, is ringed with towering sandstone cliffs and harbors a treasure trove of sea caves accessible only by kayak. Go with Trek and Trail (www.trek-trail.com, 800.354.8735). ◈

11 Niobrara River, Nebraska Not only does Nebraska have white water (surprise!), but on parts of the Niobrara, a national scenic river (www.nps.gov/niob, 402.376.1901), you might want to portage once or twice. The carry is worth it: On the 27-mile section from Cornell Bridge Launch to Last Chance Landing, outside of Valentine, you can float by waterfalls and, if you're lucky, spot grazing elk. To make a weekend of it, camp midway at Smith Falls State Park (outdoornebraska.gov/smithfalls, 402.376.1306).

12 Buffalo National River, Arkansas In early winter, temperatures tend to be moderate on Arkansas's famed Buffalo River, its signature limestone bluffs are in full view, and summer boaters are long gone (www.nps.gov/buff, 870.439.2502). Nearby, the Buffalo Outdoor Center (www.buffaloriver.com, 870.861.5514) offers rental cabins as well as single-day and multi-day canoe rentals along Ponca Creek, a tributary to the Buffalo River. ◈

WINTER SPORTS
Wolf tracks to Nordic trails

13 Grand Marais, Minnesota Throughout the winter, track timber and gray wolves with local naturalists at the Northwoods' Gunflint Lodge, located about three hours from Duluth (www.gunflint.com, 800.328.3325). You'll identify prints and learn about the wolves' social hierarchy, all while exploring pine forests and white-cedar swamps.

14 Boundary Waters Canoe Area Wilderness, Minnesota The hushed and wolf-haunted woods of the Boundary Waters are a musher's paradise during the winter. Arctic explorer Paul Schurke's Wintergreen Dogsled Lodge (www.dogsledding.com, 218.349.6128) operates multiday camping and lodge-based trips near Ely. ◈

15 Knife River Indian Villages National Historic Site, North Dakota Winter is not for the faint of heart on the Great Plains, but a visit to Knife River makes it that much more bearable (www.nps.gov/knri, 701.745.3300). The park's one-and-a-half-mile snowshoe trail leads to the remains of historic Hidatsa villages. The 11-mile network of cross-country ski trails is excellent for wildlife viewing. If you're lucky, you'll spot coyotes, deer, and occasionally antelope.

16 Cable, Wisconsin Wisconsin's Northwoods is like Norway without the fjords. The town of Cable is home to the Birkebeiner, the country's best-known cross-country ski race (www.birkie.com, 715.634.5025), and it lies at the center of 60 miles of trails. Contact New Moon Ski & Bike (www.newmoonski.com, 800.754.8685) in nearby Hayward to rent skis and for advice on which runs to try first.

Cable

◈ National Geographic topographic map available for this area.
800-962-1643
www.natgeomaps.com

BIKING
Northwoods loops to Texas Hill Country

Mississippi River Trail

17 Root River State Trail, Minnesota On the Root River State Trail in southeast Minnesota, cyclists come for pristine hardwood forests, impressive limestone bluffs, and 42 miles of road-bike-friendly asphalt. Base yourself at Cedar Valley Resort , located outside Lanesboro (www.cedarvalleyresort.com, 507.467.9000). Bike all day on an out-and-back to either trailhead. If you've got a second day to spare, go the other direction—this time on a rented recumbent bicycle from the Little River General Store (www.facebook.com/LittleRiverGeneral Store, 507.467.2943).

18 Chequamegon-Nicolet National Forest, Wisconsin Just a half day's drive from Minneapolis, Madison, or Milwaukee, the 1.5 million-acre Chequamegon-Nicolet National Forest (www.fs.usda.gov/cnnf, 715.362.1300) offers over 300 miles of bikable trails. Try the pine-lined Esker Trail (wild blueberries abound in late summer) or the Ojibwe Trail. Contact the Chequamegon Area Mountain Bike Association (www.cambatrails.org, 715.798.3599).

19 Kettle Moraine State Forest, Wisconsin The southern unit of this thick-canopied forest, an hour from Milwaukee, boasts 30 miles of mountain bike trails. For roller-coaster riding at its finest, with loops and one-way routing, check out the John Muir trail system. Contact the forest for more details (dnr.wisconsin.gov/topic/parks/kms, 262.594.6200).

20 Mississippi River Trail, Minnesota to Louisiana Finally, a bike trail to befit the nation's greatest river. The Mississippi River Trail winds through parks and wildlife refuges in ten states and across nearly 3,000 miles—from Lake Itasca, Minnesota, to the Gulf of Mexico. The loose network of state and county roads is good for an afternoon—or a lifetime—of pedaling. Search for local trails in the network by contacting state natural resources agencies, or see American Trails (www.americantrails.org).

21 X Bar Ranch, Texas Cowboy meets hiker with happy results at the X Bar Ranch (www.xbarranch.com, 325.853.2688), a 5,400-acre working cattle ranch three hours west of Austin. Three miles of nature trails wind through the hills and valleys where the Texas Hill Country meets West Texas. Terrain varies from rocky limestone outcrops to grass and oak savannas, with plenty of scenic vistas and year-round opportunities for nature-watching.

BIRDING
Whooping cranes to spruce grouse

22 Aransas National Wildlife Refuge, Texas Aransas hosts one of the few remaining populations of the endangered whooping crane, once common from Florida to Mexico. The 500 or so cranes winter here from mid-October to mid-April. Though the interior of the refuge is closed at that time, visitors can pull out the binoculars and view the birds from a network of observation towers, or they can book a birding boat tour. Contact the Aransas National Wildlife Refuge (www.fws.gov/refuge/Aransas, 361.349.1181) for information and to find boat tour operators.

Texas Gulf Coast

23 Cheyenne Bottoms Wildlife Area, Kansas The nearly 20,000-acre Cheyenne Bottoms Wildlife Area is part of the largest marsh in the nation's interior and is one of the most important migration areas in the Western Hemisphere. An large percentage of North America's shorebird population stops here during the spring. You should do the same: Come from April to May for a show of long-billed dowitchers, Wilson's phalaropes, and white-rumped, semi-palmated, and Baird's sandpipers. Contact the Great Bend Convention and Visitors Bureau (www.exploregreatbend.com, 620.792.2750) for details.

24 Dale Bumpers White River National Wildlife Refuge, Arkansas As one of the premier waterfowl wintering sites in the U.S., the 160,000-acre Dale Bumpers White River National Wildlife Refuge is not only a must-see for the more than one million wintering ducks and geese, but for numerous breeding songbirds too. For the best visit, avoid deep winter's high water and come in the spring and fall shoulder seasons. Contact the refuge (www.fws.gov/refuge/white_river/, 870.282.8200) for water levels and migrant activity.

25 Superior National Forest, Minnesota In late spring, Superior National Forest becomes one of the most bird species-rich places on the continent. View more than 220 species of neotropical migrants, including the golden-winged, black-throated blue, and Canada warbler. Or come anytime to spot year-round boreal and great gray owls, spruce grouse, three-toed woodpeckers, and gray jays. For additional details, contact the forest (www.fs.usda.gov/superior, 218.626.4300). ◈

MOUNTAIN

It was here, along the spine of the continent, that things started getting, well, rocky for Lewis and Clark. Today, however, the hundred individual ranges of the Rocky Mountains reward adventure-seekers (the ones with cars, not mules) with endless possibilities. In Colorado alone, 53 peaks soar above 14,000 feet, harboring an uncountable number of hiking, biking, skiing, and climbing destinations. The remote peaks of Montana and Idaho host equally wild adventures, and farther south, the red-rock country of Arizona, Utah, and New Mexico is riddled with wildly eroded rock formations, slot canyons, and Native American archaeological sites. The Rockies contain some of the largest tracts of protected wilderness in the lower 48, meaning much of the region's backcountry is as pristine as it was when the Corps of Discovery first beheld it. This jagged swath of North America is also home to some of our nation's most iconic natural sites: the Grand Canyon, Yellowstone, the Tetons. See them, and then aim your wheels farther into the high country. Below, a definitive selection of the adventures that await you.

Grand Canyon

HIKING AND BACKPACKING
Canyon deeps to secret cirques

1 Bob Marshall Wilderness, Montana
The "Bob" makes other wildlands look like urban centers. Include the adjoining Scapegoat and Great Bear Wildernesses and you've got more than 1.5 million acres, 1,700 miles of trails, four large river forks, 70 miles of Continental Divide Trail, and nearly every major Rocky Mountain critter. Explore Scapegoat Mountain for stunning geology and zero crowds. The Bob Marshall Wilderness Foundation has information (www.bmwf.org, 406.387.3822). ◈

2 Cirque of the Towers, Wyoming Begin the most classic hike in the Wind River Range at the Big Sandy Opening trailhead. Pack in eight miles and make camp in the cirque—a wonderland ringed by soaring tabletops, fins, and spikes of gray granite. You'll be smack in the middle of an award-worthy calendar photo. For info, contact the Pinedale Travel and Tourism Commission (www.visitpinedale.org, 888.285.7282).

3 Weminuche Wilderness Area, Colorado
The best mountain backpacking trip in the Four Corners region pairs the high-peaks terrain of the San Juan Mountains with a unique shuttle system: the Durango & Silverton Narrow Gauge Railroad. For information on the 35-mile "train loop," contact the San Juan National Forest (www.fs.usda.gov/sanjuan, 970.247.4874) and the Durango & Silverton Narrow Gauge Railroad (www.durangotrain.com, 877.872.4607), which drops hikers at the trailhead. ◈

4 Grand Canyon National Park, Arizona Six thousand feet deep and 277 miles long, it's only the Earth's most spectacular hole in the ground. Don't be like the millions of Griswolds who remain at the rim each year; real beauty—as well as 1.7 billion years of geology—lies within. Treat yourself to the show with an overnight hike to Horseshoe Mesa (and don't forget the water!). Contact the park (www.nps.gov/grca, 928.638.7888) for trail information. ◈

CLIMBING
Insiders' Tetons to rock city

5 City of Rocks National Reserve, Idaho
The spectacular granite towers of the City of Rocks were, by the middle of the 19th century, a landmark for pioneers heading west on the California Trail. Today's pioneers are climbers eager for ascents. Join Sawtooth Mountain Guides (www.sawtoothguides.com, 208.806.3063) for a day of cruising the area's hundreds of established routes.

6 Mount Owen and Mount Moran, Wyoming The obvious choice for most mountaineers in the Teton Range is 13,770-foot Grand Teton, but guides prefer empty and dramatic Mount Owen (12,928 feet) and Mount Moran (12,605 feet); both are multiday climbs. Contact Exum Mountain Guides (exumguides.com, 307.733.2297) to arrange a trip. ◈

7 Eldorado Canyon State Park, Colorado
This popular spot near Boulder harbors hundreds of trad climbs with legendarily hard-to-place protection. The Bastille Crack (5.7), a classic, runs up the most prominent buttress, demanding a mix of face moves, stemming, and perfect hand jams. Locals head to Over the Hill (5.10b) up a left-facing corner. Contact Colorado Mountain School (coloradomountainschool.com, 720.387.8944) for details. ◈

8 Castleton Tower, Utah Of all the mighty monoliths on the Colorado Plateau, the climbers' favorite is Castleton Tower, which combines great rock, challenging routes, and a stunning position atop a 1,200-foot cone of scree. Fortunately for nonexpert climbers, two of the best climbs are also relatively moderate: The four-pitch North Chimney and Kor-Ingalls routes are both 5.9. For info, contact Moab Desert Adventures (www.moabdesertadventures.com, 804.814.3872) and the Bureau of Land Management, Moab Field Office (www.blm.gov/office/moab-field-office, 435.259.2100). ◈

Moab

◆ National Geographic topographic map available for this area.
800-962-1643
www.natgeomaps.com

PADDLING
Desert floats to spring torrents

9 Middle Fork of the Salmon River, Idaho Outdoor Adventure River Specialists (www.oars.com, 800.346.6277) runs six-day trips down Idaho's Middle Fork of the Salmon, which tumbles through a hundred rapids in a hundred miles. Water levels permitting, you'll put in at Indian Creek, then plunge a heart-stopping 41 feet per mile during the first 25 miles. The remainder surfs and spins through dense forests, grassy meadows, and imposing canyons until the take-out at the main Salmon.

10 Arkansas River, Colorado The Upper Arkansas River plunges through some of North America's most spectacular mountain scenery—not that paddlers have time to take their eyes off the water. Between Leadville and Salida, the Arkansas churns in a nearly continuous chain of Class III–V rapids. Famed sections like Royal Gorge, Browns Canyons, and the Numbers can be run individually or together on a trip with Rocky Mountain Outdoor Center rmoc.com, 719.395.3335). ◆

11 Verde River, Arizona Each year, Arizona's spring thaw boosts the sluggish Verde River into a 170-mile torrent that runs from Flagstaff to Phoenix. Kayakers take to the Class II–IV rapids on the two-day, 17-mile section between Beasley Flat and Childs Campground, located a hundred miles from Phoenix. Canoeists typically opt for the gentler three-day, 42-mile run from Childs Campground to Horseshoe Dam (www.visitarizona.com, 866.275.5816). Verde Adventures offers guided kayak tours (www.sedonaadventuretours.com, 877.673.3661). ◆

Verde River

12 Rio Chama, New Mexico One of New Mexico's best rafting trips—the three-day journey down the Rio Chama—begins atop a high plateau and descends through the multicolored sandstone world immortalized in the paintings of Georgia O'Keeffe. Far Flung Adventures (www.farflung.com, 575.758.2628) runs trips.

WINTER SPORTS
Backcountry forays to big-time resort runs

13 Sun Valley, Idaho "Sun Valley is the best mountain in the world," the late ski bum and filmmaker Warren Miller once said. The resort has grown considerably over the years, with 18 lifts, 25 miles of trails, and 2,434 skiable acres. But the Bald Mountain area, where Miller skied, is still home to Sun Valley's most exciting runs. Contact the Sun Valley Resort (www.sunvalley.com, 800.786.8259).

14 Jackson Hole, Wyoming The Tetons are the epicenter of American ski mountaineering, with the Grand still holding its own as one of the most serious descents anywhere. Yet Teton Pass and the nearby Jackson Hole ski area serve up freshies manageable by even a first-time adventurer. Contact The Mountain Guides (themountainguides. com, 800.239.7642) for guided trips. ◆

Ouray

15 Ouray, Colorado Ice climbing may never achieve the mass appeal of, say, snowboarding, but the Ouray Ice Park in southwestern Colorado has gone a long way toward popularizing the sport (www.ourayicepark. com, 970.325.4288). The park has over 100 man-made routes, ranging from 40 to 150 feet high, all within a mile-long section of Uncompahgre Gorge. San Juan Mountain Guides offers full-day private lessons and a variety of longer clinics (mtnguide.net, 800.642.5389). ◆

16 La Sal Mountains, Utah In the La Sal Mountain Hut System, backcountry skiers and snowboarders can stay in a series of huts high in the 10,000-foot, pine- and aspen-clad mountains above Moab. Reachable via snowcat trails, the huts access bowls, glades, and a network of Nordic ski trails, all buried in Utah's finest powder. Contact Ski Utah (www. skiutah.com/snowreport, 800.754.8824) for current snow conditions. ◆

BIKING
Grassland tracks to slickrock grail trails

17 Ketchum, Idaho There are literally thousands of miles of bikable trail near Ketchum—a daunting number for an outsider. Start with a known quantity. Sun Valley's seven-mile Cold Springs Trail and the five-mile Warm Springs Trail (both lift-served) wind through Douglas firs and bomb down sage-covered slopes, and the Adams Gulch Loop spins seven miles through aspen groves just outside of town. Contact Sturtevants (sturtevants-sv.com, 208.726.4501) for bike rentals and info.

18 Moab, Utah Believe the hype: Moab has phenomenal singletrack, challenging slickrock, and some of the most spectacular vistas on the planet. First-time visitors should ride the classic Slickrock Trail once, then hit the real crown jewel—Porcupine Rim, 15 miles of singletrack, rock gardens, and sensational views, followed by six miles of road. Poison Spider Bicycles is a good source of local info (poisonspiderbicycles.com, 800.635.1792). ◆

19 Comanche National Grassland, Colorado Bring a bike and head 90 miles east of Pueblo to the rugged doubletrack of Picket Wire Canyonlands in Comanche National Grassland (www.fs.usda.gov/psicc, 719.533.1400). The 20-mile ride takes in short-grass prairie, red-rock canyons—and over 1,900 150-million-year-old dinosaur footprints in 130 separate trackways.

20 Sedona, Arizona Sedona is perhaps better known for spas than for sports, but the town—set in the heart of the red-rock West and surrounded by a thousand miles of trails—is a premier base camp for mountain biking and hiking. Go in November to tackle routes that were frying-pan hot a month earlier. Bike & Bean rents bikes (www.bike-bean.com, 928.284.0210). ◆

BIRDING
Black swifts to flame-colored tanagers

21 Glacier National Park, Montana With 700 lakes and numerous mountain streams, Glacier supports a vast range of high-elevation habitats for breeding waterfowl, hawks, owls, and woodpeckers. Access is limited in winter months, so come in the late spring or early fall, binoculars in hand, to ply the park's 700 miles of trails without the logjam of summer crowds. Contact the park for the latest weather and trail conditions (www.nps.gov/glac, 406.888.7800). ◆

22 Great Salt Lake, Utah The rich larder of brine shrimp and brine flies in Great Salt Lake attracts vast flocks of migrating waterfowl and shorebirds. Of particular interest to birders is the rare snowy plover. There are a number of protected areas around the lake, but the best one for birding is Antelope Island State Park (stateparks.utah.gov/parks/antelope-island, 801.773.2941).

23 Rocky Mountain National Park, Colorado It so happens that the best birding areas in Rocky Mountain National Park are also some of the hardest to get to. To reach the best coniferous forests and alpine grasslands within the park's wilderness areas, you'll need fortitude, but in return you'll get an extraordinary display of rare black swifts and high-elevation brown-capped rosy-finches, with few if any of the park's over four million annual visitors as a distraction. To learn more, contact Rocky Mountain National Park (www.nps.gov/romo, 970.586.1206). ◆

24 Bosque del Apache National Wildlife Refuge, New Mexico A glimpse of an endangered sandhill crane, tundra swan, white pelican, or snow goose is enough incentive to pay a visit to Bosque del Apache, a stark desert landscape nestled between barren mountains. Add to that the possibility of seeing an endangered whooping crane or a rare trumpeter swan, plus a host of songbirds, and you have a real birding mecca. Contact the refuge for info (www.fws.gov/refuge/bosque_del_apache/, 575.835.1828).

Bosque del Apache

25 Madera Canyon, Arizona For some of Mexico's best and brightest, check out southern Arizona's Madera Canyon. Birders come from around the globe to scope out the dazzling display of neotropical migrants and rarities such as Mexican jays and eared trogons. Check with locals if you plan to probe the canyon environs, as road wash-outs are common. Friends of Madera Canyon (www.friendsofmaderacanyon.org) doles out good advice.

PACIFIC

This is where it all ends, and where it all begins.
The Pacific is the end of the line for westbound road-trippers, and it is the birthplace of innovation for adventure sports. The Pacific region's staggering variety of terrain and eco-zones—alpine, temperate and tropical rainforest, tundra, desert—makes it a bona fide promised land. Backpackers tackle the legendary border-to-border Pacific Crest Trail. Peak-baggers climb to their heart's content in the Sierra Nevada and Cascades. Skiers and snow-boarders enjoy a smorgasbord of groomed and out-of-bounds runs in every imaginable setting. And with no shortage of water—salt or fresh—paddlers can pick and choose as they please: from sea kayaking Washington's San Juan Islands to running Class IV-plus rapids on the Klamath in northern California and Oregon. Conservationist John Muir once wrote of his beloved Sierra, "New beauty meets us at every step in all our wanderings." Was there ever any better inspiration for a road trip? Below, some of the Far West's best destinations.

Death Valley

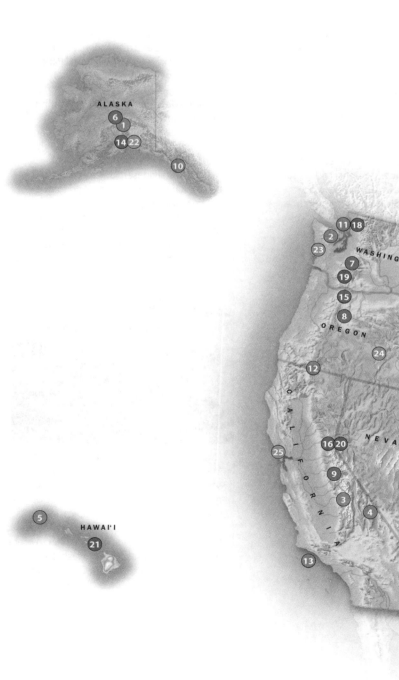

HIKING AND BACKPACKING
California's high country to Hawaiian rainforest

1 Denali State Park, Alaska For tremendous views of Denali (Mt. McKinley), hike the Kesugi Ridge Trail in often overlooked Denali State Park. Start at Little Coal Creek trailhead to get above timberline in a hurry. The long ridge is easy to follow; the traverse takes most hikers three days. Contact Denali State Park for details (dnr.alaska.gov/parks/units/denali1.htm, 907.745.3975). ◈

2 Olympic National Park, Washington For one of the most remote coastal experiences in the nation, tackle Olympic National Park's Wilderness Coast. Covering 73 miles of untamed beach (which can be done in sections), you'll scramble up rocky headlands, marvel at stands of oversize Sitka spruce, and search the tidepools for wildly colored starfish and anemones. Contact the park for information (www.nps.gov/olym, 360.565.3130). ◈

3 Sequoia and Kings Canyon National Park, California The southern Sierra is home to the largest trees on Earth, the highest point in the lower 48, and some of the best backpacking trails anywhere. Insiders head to the lightly visited Mineral King area, a warren of lakes and valleys surrounded by the peaks of the Great Western Divide, all guarded by a 25-mile road with 698 curves. Contact the park for directions (www.nps.gov/seki, 559.565.3341). ◈

4 Death Valley National Park, California Extreme heat makes Death Valley famous, but extreme height—11,331 feet, desert floor to mountaintop—is what makes it compelling. The seven-mile route to Telescope Peak (11,049 feet) begins in pine forest at the Mahogany Flat campground but quickly reaches the open crest of the Panamint Range, where vistas open out across thousands of square miles. Contact the park for info (www.nps.gov/deva, 760.786.3200). ◈

5 Nāpali Coast State Wilderness Park, Hawai'i Coastal hikers generally agree that the 11-mile Kalalau Trail edging along Kauai's Nāpali Coast is the crown jewel of beach hikes. But that doesn't mean it's a cakewalk. Be ready to ford streams, watch tides, and camp in pristine jungle. The trail meanders from cliff to shore before spilling out at remote and spectacular Kalalau Beach. The Division of State Parks issues permits (dlnr.hawaii.gov/dsp/parks/kauai/napali-coast-state-wilderness-park, 808.274.3444).

CLIMBING
Big-wall Yosemite to backcountry Alaska

6 Denali National Park and Preserve, Alaska Alaska Mountaineering School (www.climbalaska.org, 907.733.1016) offers an inside edge on building off-trail exploration skills with its weeklong wilderness courses deep into the seldom visited area south of the Alaska Range. The trip also features a climb up an unnamed peak or two. More than 20 hours of available summer daylight means leisurely evenings spent scanning for moose, grizzlies, and elusive wolves. ◈

7 Mount Rainier, Washington The perfect proving ground for fledgling mountaineers, Rainier's classic Disappointment Cleaver route takes climbers up a 1,600-foot snowfield, across three glaciers, and onto the broad 14,410-foot summit. Rainier Mountaineering, Inc. (rmiguides.com, 888.892.5462) offers acclaimed four- and five-day climbing courses that conclude with a summit attempt. ◈

8 Smith Rock State Park, Oregon Climbers in central Oregon's Smith Rock State Park (stateparks.oregon.gov, 541.548.7501) have mapped over 1,000 bolted routes up the cliffs lining the Crooked River. Pick one of the countless formations—Christian Brothers, Four Horsemen, or the alarmingly lifelike Monkey Face—and go. Timberline Mountain Guides offers climbing classes (timberlinemtguides.com, 541.312.9242). ◈

9 Yosemite National Park, California Yosemite was the birthplace of American big-wall climbing, and top climbers from around the world still come here to test themselves. Try an all-time classic: For more advanced climbers, the East Buttress of Middle Cathedral (5.10c) is 11 pitches of dreamy golden granite. Yosemite Mountaineering School (www.travelyosemite.com, 209.372.8344) has guides. ◈

WINTER SPORTS
Island backcountry to terrain-park thrills

14 Chugach Mountains, Alaska The Tsaina Lodge (valdezheliskiguides.com/tsaina-lodge, 907.835.4528) atop Thompson Pass north of Valdez is a modern, contemporary-styled ski lodge surrounded by some of the most breathtaking skiing available anywhere. In the midst of the 5.4-million-acre Chugach National Forest, mild, wide-open cruisers exist happily alongside alleyway chutes. The area near the lodge gets about 200 inches of annual snowfall; after hiking up a couple of hours, you're in 500-plus territory.

Chugach Mountains

PADDLING
Island hopping to easy-access white water

10 Glacier Bay National Park and Preserve, Alaska Glacier Bay National Park and Preserve—with its calving ice masses and breaching 30-ton whales—is a sensory theater without equal, and the cockpit of a sea kayak is the best seat in the house. Intermediate boaters can tackle these protected waters on their own; less experienced paddlers can join a guided day trip. Glacier Bay Sea Kayaks (glacierbayseakayaks.com, 907.697.2257), the concessionaire for the park, has boats and guides.

11 San Juan Islands, Washington To duck the crowds that flock to Puget Sound's San Juan Islands in the summer and early fall, decamp for sleepy Shaw Island. From sandy Shaw Island County Park, launch on a day-long mission to circle Shaw or explore nearby Lopez and Turn Islands, where you can watch for bald eagles. Contact San Juan County Parks (www.sanjuanco.com/523/Shaw-Island, 360.378.8420).

San Juan Islands

12 Upper Klamath River, Oregon/California Rafting guides running the Hell's Corner Gorge on the Upper Klamath River don't bother with the usual banter—no one would hear it. The river does all the talking. In the gorge's heart, the Class IV-plus Klamath knifes deep into the basalt core of the Cascade Range and rafters collide with 30 rapids in 17 miles. River Dancers (www.riverdancers.com, 530.918.8610), based in Mount Shasta, California, runs the river in trips of one to five days.

13 Channel Islands National Park, California Santa Cruz Island, the largest of the five undeveloped isles of Channel Islands National Park (www.nps.gov/chis, 805.658.5730), has hundreds of sea caves in its fortresslike northern cliffs. Experienced paddlers can ferry to the island from Ventura or Oxnard and begin exploring the caverns; go unguided or hook up with an outfitter such as Channel Islands Adventure Company (www.islandkayaking.com, 805.884.9283).

15 Mount Hood, Oregon For lift-served skiing in both the dead of winter and the dog days of summer, snowhounds in the lower 48 have only one option: Mount Hood. The perennial Palmer Glacier hosts squads of snowboarders and ski racers who come to train on the long, 30-degree runs. The Timberline Lodge (www.timberlinelodge.com, 503.272.3311) is the hub for all mountain activity, summer or winter.

16 Lake Tahoe, California Tahoe is the unsung hero of lift-accessed out-of-bounds skiing. Northstar has an open-boundary policy, while Alpine Meadows and Sierra offer guided tours into the volcanic moonscape and remote valleys far from the bunny-slope groomers (www.skilaketahoe.com and www.northstarcalifornia.com, 800.466.6784).

17 Ruby Mountains, Nevada The snow in northeastern Nevada, like that of Utah, is dry, with snow-to-liquid ratios well over 10:1. But unlike Utah, there are no major ski resorts; Nevada's high country, most notably the Ruby Mountains, is empty and wild. To explore the glacial cirques and glades of white pine, skiers can spend days hiking in or get there in minutes on a helicopter. You choose. Contact Ruby Mountain Helicopter Experience for information (www.helicopterskiing.com, 775.753.6867).

BIKING
Lunar landscapes to singletrack Shangri-la

18 Bellingham, Washington For the perfect tour of one of the Northwest's hottest biking destinations, start on Chuckanut Mountain outside Bellingham, whose singletrack-laced ridgelines provide views of the San Juan Islands. If you're still feeling peppy, nearby Galbraith has 80-plus trails. Camp at Larrabee State Park (parks.state.wa.us/537/Larrabee, 888.226.7688) and rent your ride from Jack's Bicycle Center (www.jacksbicyclecenter.com, 360.733.1955).

19 Mount St. Helens, Washington Still think of Mount St. Helens as that dusty, fuming cone that blew up in 1980? Time to reacquaint yourself. Take a ride down the five-and-a-half-mile Ape Canyon Trail and you'll sail on rollicking singletrack from mudflows and remnants of old-growth forest to meadows and lunar plains. Contact the Mount St. Helens National Volcanic Monument in Gifford Pinchot National Forest for details and trail conditions (www.fs.usda.gov/giffordpinchot, 360.449.7800).

20 Lake Tahoe, California/Nevada Mountain bikers flock to the justly famous Flume Trail on the lake's east shore. For challenge without company, steer to tricky Mr. Toad's Wild Ride or the thigh-burning Tahoe Rim Trail, at the point where it runs south from Heavenly ski area toward Armstrong Pass. For bikes and maps: www.tahoesbest.com/biking; www.tahoebike.org, 775.298.0273.

21 Haleakalā National Park, Hawai'i Drive up Crater Road, which switchbacks from sea level to nearly 10,000 feet. Hundreds of tourists road bike from the top every day, but you're taking the Skyline Trail for a wilder way down: ultimately dropping 6,500 feet in 13 thrilling miles. Contact Haleakalā National Park (www.nps.gov/hale, 808.572.4400). Bike Maui runs trips in the park (www.bikemaui.com, 808.575.9575).

BIRDING
Shorebirds to desert dwellers

22 Copper River Delta, Alaska A 700,000-acre string of wetlands, islands, and mud-flats at the base of the Chugach Mountains, the Copper River Delta is the largest wetland on the Pacific Coast, drawing some 12 million migrating shorebirds annually—more than anywhere else in the Western Hemisphere. Each May, folks in the local town of Cordova take to the streets for the annual Copper River Delta Shorebird Festival (cordovachamber.com, 907.424.7260), with workshops and events to celebrate their unique birding status. Participants learn the ways of western sandpipers, trumpeter swans, dusky Canada geese, and a host of other visiting migrants. For more information contact Chugach National Forest (www.fs.usda.gov/chugach, 907.743.9500).

23 Grays Harbor, Washington To put it plainly, when hundreds of thousands of shorebirds show up at Grays Harbor, on the state's Pacific coast, it's a pig-out. Scores of dunlins, short-billed and long-billed dowitchers, and semipalmated plovers launch into a weight-doubling feeding frenzy before their flight to their summer grounds in Alaska. Visit between April and May. Contact the Billy Frank Jr. Nisqually National Wildlife Refuge (www.fws.gov/refuge/Billy_Frank_Jr_Nisqually, 360.753.9467) to learn more.

24 Malheur National Wildlife Refuge, Oregon In 1908 the high desert refuge of Malheur was set aside to protect nesting herons, egrets, and ibis from plume hunters. Thank goodness. Today, scope-toting visitorsand photographers come to stalk the huge populations of wading birds and perennial populations of the short-eared owl, greater sage-grouse, and bobolink. Avoid the summer mosquitoes; come in spring or fall. The refuge has information (www.fws.gov/refuge/malheur/, 541.493.2612).

25 Point Reyes National Seashore, California Some of the best birding in the country happens to be only 20 miles north of San Francisco. At Point Blue Conservation Science (www.pointblue.org, 707.781.2555), in the heart of the national seashore, you'll find resident snow plovers, black oystercatchers, and, if you're lucky, a small population of California spotted owls. The trained staff biologists will answer questions and occasionally run outings (www.nps.gov/pore, 415.464.5100 x2).

Lake Tahoe

National Geographic topographic map available for this area.
800-962-1643
www.natgeomaps.com

MI 100 200 300 400 500
KM 100 200 300 400 500

The National Park System protects areas of natural and historical importance in the U.S., ranging from scenic rivers to historic battlefields. In addition to the 63 national parks, the park system maintains sites as diverse as a cattle ranch in Montana and Independence Hall in Philadelphia.

This map shows national parks as well as other areas protected by the National Park Service. National Park names are shown with bold type.

Abbreviations

IHS	International Historic Site
NB	National Battlefield
NBP	National Battlefield Park
NBS	National Battlefield Site
NHEA	National Heritage Area
NHA	National Historic Area
NHP	National Historical Park
NHP & PRES	National Historical Park & Preserve
NH RES	National Historical Reserve
NHS	National Historic Site
NL	National Lakeshore
NM	National Monument
NM & PRES	National Monument & Preserve
NMP	National Military Park
N MEM	National Memorial
NP	National Park
NP & PRES	National Park & Preserve
N PRES	National Preserve
NR	National River
NRA	National Recreation Area
NRR	National Recreational River
NRRA	National River & Recreation Area
N RES	National Reserve
NS	National Seashore
NSR	National Scenic River/Riverway
NST	National Scenic Trail
PKWY	Parkway
SRR	Scenic and Recreational River
WR	Wild River
WSR	Wild & Scenic River

Areas other than national parks that are protected by the National Park System:

San Francisco Area

Eugene O'Neill NHS
Fort Point NHS
Golden Gate NRA
John Muir NHS
Port Chicago Naval
  Magazine N MEM
Rosie the Riveter/
  WWII Home Front NHP
San Francisco
  Maritime NHP

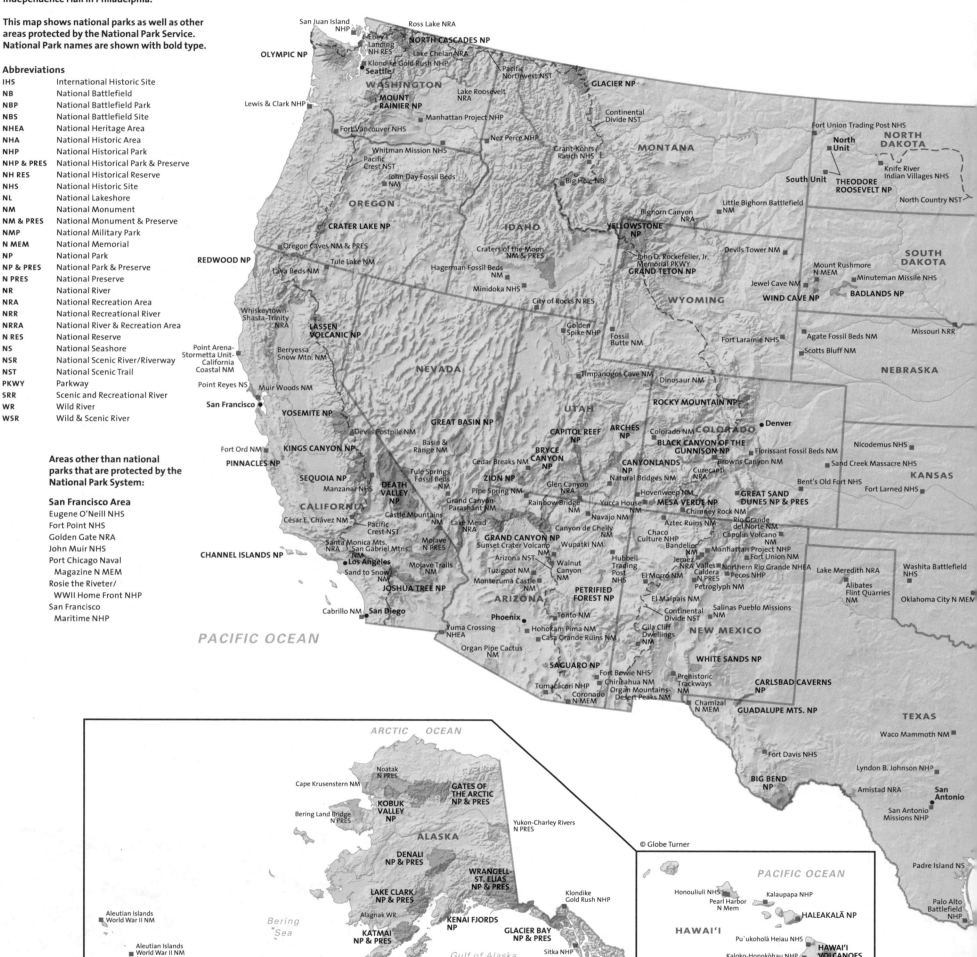

Boston Area
Adams NHP
Boston African American NHS
Boston Harbor Islands NRA
Boston NHP
Frederick Law Olmsted NHS
John F. Kennedy NHS
Longfellow House -
 Washington's Headquarters NHS
Minute Man NHP
Salem Maritime NHS
Saugus Iron Works NHS

New York City Area
African Burial Ground NM
Castle Clinton NM
Federal Hall N MEM
General Grant N MEM
Governors Island NM
Hamilton Grange N MEM
Lower East Side Tenement Museum NHS
National Parks of New York Harbor
Sagamore Hill NHS
Saint Paul's Church NHS
Statue of Liberty NM
Stonewall NM
Theodore Roosevelt Birthplace NHS
Thomas Cole NHS
Thomas Edison NHP

Philadelphia Area
Edgar Allan Poe NHS
Gloria Dei Church NHS
Independence NHP
Thaddeus Kosciuszko N MEM

Delaware
First State NHP

Baltimore Area
Baltimore-Washington PKWY
Ft. McHenry NM and Historic Shrine
Hampton NHS

District of Columbia
African American Civil War Memorial NM
Anacostia Park
Belmont-Paul Women's Equality NM
Capitol Hill Parks
Carter G. Woodson Home NHS
Constitution Gardens
Ford's Theatre NHS
Fort Dupont Park
Franklin Delano Roosevelt Memorial
Frederick Douglass NHS
George Mason Memorial
John Ericsson N MEM
Kenilworth Park & Aquatic Gardens
Korean War Veterans Memorial
LBJ Memorial Grove
Lincoln Memorial
Martin Luther King, Jr. Memorial
Mary McLeod Bethune Council House NHS
Meridian Hill Park
National Capital Parks-East
National Mall and Memorial Parks
National WWII Memorial
Old Post Office Tower
Peirce Mill
Pennsylvania Avenue NHS
Rock Creek Park
The Old Stone House
Theodore Roosevelt Island
Thomas Jefferson Memorial
Vietnam Veterans Memorial
Washington Monument
White House

Maryland
Chesapeake and Ohio Canal NHP
Clara Barton NHS
Fort McHenry NM and Historic Shrine
Fort Washington Park
Greenbelt Park
Hampton NHS
Harmony Hall
Harriet Tubman Underground R.R. NHP
Monocacy NB
Oxen Cove Park & Oxen Hill Farm
Piscataway Park
Potomac Heritage NST

Virginia
Arlington House
George Washington Memorial PKWY
Great Falls Park
Wolf Trap NP for the Performing Arts

Guam
War in the Pacific NHP

American Samoa
National Park of American Samoa

ACADIA

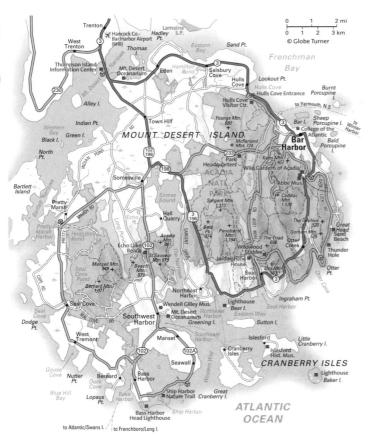

© Globe Turner

Central coastal Maine
Established February 26, 1919
49,077 acres

Headquarters
Acadia National Park
PO Box 177
Bar Harbor, ME 04609
207.288.3338
www.nps.gov/acad
@AcadiaNPS

Visitor Centers
Hulls Cove Visitor Center
(207.288.3338), on Rte. 3 just
before entrance to Park Loop Rd.,
open early May to late October.
Sieur de Monts Nature Center,
Rte. 3 south of Bar Harbor, open
mid-May to mid-October. Village
Green Information Center, Bar
Harbor, May to October.

Entrance Fees
$30 per vehicle, $25 per motorcycle, $15 per person on foot or
bicycle, all valid for 7 days.

Accommodations
Two campgrounds on Mt. Desert
Island. Blackwoods, open early
May to mid-Oct, reservations
required. Seawall, open late May
to mid-Oct, reservations required.
Five primitive shelters at Duck
Harbor Campground, on Isle au
Haut, are available by reservation
from mid-May to mid-Oct. Nearby
private campgrounds offer addi-
tional sites, mainly May to Oct.

When to Go
Like all of New England, the
Maine coast has notoriously fickle
weather. During summer, the
most popular time to visit,
temperatures seldom rise above
80°F (although traffic can cause
personal temperatures to rise
much higher), and fog is common.
The shoulder months of May
and Sept are good times to avoid
crowds and cold. For the heartier
souls who don't mind bone-chill-
ing winds and frigid temperatures,
the off-season can be a rewarding
time to visit: The park stays open
year-round, though the Park Loop
Rd., except for two short sections,
will close for inclement winter
weather.

For its nearly three million
visitors each year, Acadia National Park *is* Maine,
a place where evergreen
forests and stark granite
cliffs meet crashing surf and
teeming tidal pools. The park
covers about half of Mount
Desert Island, a rocky bite of
land just off Maine's coast,
north of Penobscot Bay.
About five times larger than
Manhattan, the island
is divided into two lobes
by Somes Sound, the only
true fjord in the eastern U.S.
Mount Desert features five
large lakes and more than a
dozen mountains, including
Cadillac Mountain, whose
1,530-foot summit overlooking Frenchman Bay is the
highest Atlantic coastal
point north of Rio de Janeiro.
It boasts peaks of granite
and plentiful blueberry
bushes.

The Wabanaki called the
rocky, mountainous island
Pemetic, "the sloping land."
In 1604, the peripatetic
French explorer Samuel de
Champlain spotted smoke
from Wabanaki campfires
and, pulling closer, ran ashore
near Otter Point. Champlain
called it the "Isles des Monts
Desert" – Island of Barren
Mountains. Nine years later,
Jesuits established the first
French mission in America
on what is now Fernald Point,
near the entrance to Somes
Sound. The short-lived
settlement was part of the
French province of Aca-
dia, from which the park
obtained its name. Today's
busy summer retreat came
into being during the Gilded
Age, when Astors, Carnegies,
Fords, Morgans, and Vander-

Seaside cliffs in winter

bilts built summer mansions
on Mount Desert. Led by the
Boston textile millionaire
George Dorr, who would
become Acadia's first super-
intendent, the "summer
people" helped found the
park by donating land for
public use – John D. Rockefeller, Jr., built the island's
57 miles of gravel carriage
roads, banning automobiles
from them, and gave more
than 11,000 acres to the park.

The best way to see the
park is the popular – and
in summer, overcrowded
– 27-mile Park Loop Road.
Among the top attractions
on the road are Sand Beach,
a swimming beach made up
of finely ground shells and
sand (be warned that the
parking lot fills up early in
the summer); Great Head,
one of the highest sheer
Atlantic headlands in the

country; Thunder Hole, where
water rushes into a coastal
cave with a roar; and Cadillac Mountain. More than
150 miles of hiking trails
crisscross the park, connecting with the carriage roads
(45 miles of which are in the
park), and offer walks from
short, level ambles along the
ocean to grueling climbs.
A number of shops in the
touristy summer town of
Bar Harbor rent bikes to ride
on the carriage roads, which
are closed to motor vehicles,
or on Park Loop Road (which
can be a bit nerve-racking
in heavy traffic). Wildwood
Stables, near Jordan Pond,
offers a variety of horse-
drawn carriage rides on the
carriage roads from May
through October.

ARCHES

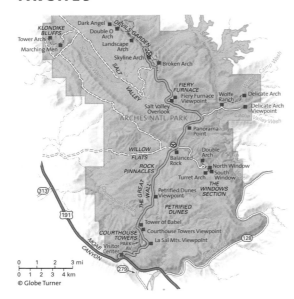

© Globe Turner

Eastern Utah
Established November 12, 1971
76,679 acres

Headquarters
Arches National Park
PO Box 907
Moab, UT 84532
435.719.2299
www.nps.gov/arch
@ArchesNPS

Visitor Center
Arches National Park Visitor
Center (435.719.2299), on U.S.
Rte. 191 at park entrance, open
daily except Christmas.

Entrance Fees
$30 per vehicle, $25 per motorcycle, $15 per person on foot or
bicycle, all valid for 7 days.

Accommodations
Devils Garden Campground,
normally open all year. For winter
reservations, call 877.444.6777. No
lodging in park. For other accommodations, contact Moab Area
Travel Council (discovermoab
.com, 435.259.8825).

When to Go
The park is in the desert, so weather conditions can be extreme, with
highs over 100°F in summer and
below 32°F in winter. Spring and
fall are the ideal times to visit,
especially for high-desert hiking.
Colorful wildflowers bloom from
April to June.

With slabs of stone rising
up like natural skyscrap-
ers, bridges and arches of
stone perched atop slick-
rock shelves, and perfectly
balanced rock piles that
resemble animals and even
humans, it's hard not to
think of Arches National Park
as a giant funland for the
gods. One can imagine titans
playing for eons in this high-
desert sandbox, arranging
and rearranging giant red
rock heaps into windows,
tables, and arches. Then one
day, the giants departed,
leaving this isolated corner
of desert as it exists today –
a natural phenomenon with
one of the largest concentra-
tions of sandstone arches in
the world. The extraordinary
spires, pinnacles, pedestals,
and stone arches have been
sculpted by wind and rain
for 150 million years. In the
Jurassic period, a 300-foot
layer of soft red sand, called
the Entrada Sandstone,
was first deposited here. As
underlying salt deposits
dissolved, it buckled and
weathered into a jumble of
slabs called fins – leading to
the more than 2,000 natural
arches seen today.

This is a visitor-friendly
park, with a number of short,
well-maintained trails that
lead directly to the major
attractions. Most visitors
are content to see the park
by taking the 18-mile scenic
drive (one way), and perhaps
settling for at least one short
hike to get a close look at one
of the arches. Such a visit can
easily be accomplished in one
day. From the visitor center,

the road climbs up from
the floor of Moab Canyon
through a slickrock expanse
known as the Petrified
Dunes, where ancient sand
berms long ago turned to
stone. Continue over the
paved road to The Windows
Section, where you can
see and photograph eight
immense arches and some
smaller formations as well.
This is also a good spot for
a short hike – just a quarter
mile to North Window. A half-
mile trail affords a dramatic
close-up of Double Arch, a
formerly solid, huge rock –
one of its openings is the
third largest in the park.

The road ends at the Dev-
ils Garden Trailhead. If you're
only going to take one hike,
this should be it. The two-
and-a-quarter-mile route

gives you views of up to 12
arches, including Landscape
Arch, a long, seemingly grav-
ity-defying thin ribbon of
stone that is considered one
of the most beautiful arches
in the park.

A spur road off the main
drive leads one and a half
miles to the remains of a
primitive cattle ranch run
by Civil War veteran John
Wesley Wolfe in the late
1800s. Nearby, petroglyphs
carved into a low cliff attest
to the presence of Native
Americans in this land, with
well-preserved images of
horse riders and what is pos-
sibly a bighorn sheep hunt.
From Wolfe Ranch, there are
two ways to see the fabulous,
much-photographed Deli-
cate Arch – by driving a mile
further to the Delicate Arch

Delicate Arch in winter

Viewpoint or by setting out
on a roughly two-hour hike
to the arch itself. Crossing
a slickrock expanse, the
trail climbs 500 feet and
offers magnificent vistas of
the desert. The arch looms
suddenly at the trail's end,
perched on a bowl-like slope
and framing white-capped
mountains in the distance.

Although the park is
essentially a desert area with
little vegetation, nearly 200
species of birds have been
spotted here, making it a
popular spot for bird-watch-
ing. Outfitters from nearby
Moab also run river trips,
jeep tours, and horseback
trips in the park vicinity.

BADLANDS

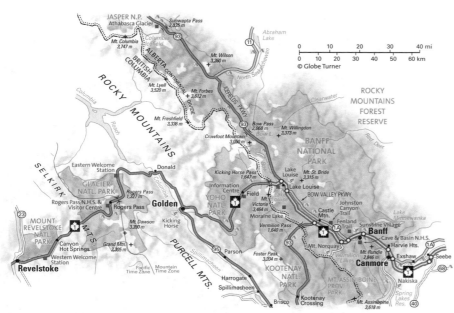

Southwestern South Dakota
Established November 10, 1978
242,756 acres

Headquarters
Badlands National Park
25216 Ben Reifel Road
Interior, SD 57750
605.433.5361
www.nps.gov/badl
@BadlandsNPS

Visitor Centers
Ben Reifel Visitor Center
(605.433.5361), at Cedar Pass on
Badlands Loop Rd./Rte. 240, open
daily except major holidays.
White River Visitor Center
(605.455.2878 in summer), in the
Stronghold Unit via Rte. 27, open
summer only.

Entrance Fees
$30 per vehicle, $25 per motorcycle,
$15 per person on foot or bicycle, all
valid for 7 days.

Accommodations
Two campgrounds, available
on a first-come, first-served
basis: Cedar Pass (open all year,
877.386.4383) and Sage Creek
(open all year, no running water
available). Cedar Pass Lodge
(877.386.4383), the only lodging
inside the park, open mid-April
through late October. For nearby
lodging outside the park, contact
the Wall Badlands Area Chamber
of Commerce (www.wall-bad
lands.com, 888.852.9255).

When to Go
Summer is most popular, but day-
time temperatures can reach 100°F,
and severe thunderstorms are not
unusual. Spring and fall are pleasant, with clear, cool days and fewer
crowds – optimum times are April
to mid-May and then Sept to mid-
Oct. In the fall, golden light colors
the canyons, and migrating birds
pass overhead. Winter is bitterly
cold. In any season, weather can
change quickly; protect against the
strong sun even in winter.

The Lakota called the ghostly
buttes and spires that rise
out of the rolling Great Plains
prairie *mako sica* – "bad land."
Lt. Col. George Armstrong
Custer called the fantastic
pinnacles and tortuous
gullies hell without the fire.
No matter how you describe
the landscape, it is unlike any-
place else on earth, thanks to
three million years of erosion
from wind and rain. Here,
where the Lakota danced
their last Ghost Dance atop
Stronghold Table in 1890, is
a harsh, arid environment,
with frequent droughts and
a winter that lasts half the
year. Though viewed as a
wasteland by most 19th-
century pioneers who passed
through quickly in their
search for farm and grazing
land, the area is nonetheless
a geologic and ecological
wonder. Sharply eroded
buttes and colorful spires
are just part of the largest
protected mixed-grass prairie
in the United States. More
than 50 kinds of wild grasses
and 200 types of wildflowers
grow in the Badlands. Below
the surface is the world's
richest repository of fossils
from the Oligocene epoch,
dating 23 to 34 million years
ago. Paleontologists have
un-earthed the remains of
sabertoothed cats, three-
toed horses, and early ances-
tors of the hog, camel, and
rhinoceros.

The easiest way to view
the park is by driving the
32-mile Badlands Loop, which
swoops down from I-90
along South Dakota Route
240 before exiting the North

Unit of the park at the Pinnacles Entrance. Just inside the
park's Northeast Entrance
is Big Badlands Overlook,
which provides your first
view of The Wall, the park's
focal point. This 100-mile-
long spine of pastel-colored
stone serves as a dramatic
backdrop for the herds of
bison, pronghorn, and bighorn sheep that wander the
lower prairie. Further along
the highway is the Windows
Overlook, which serves as
the trailhead for three short
nature trails – the Door, Window, and Notch Trails. Door
Trail is an easy three-quarter-
mile walk leading through
a natural doorway in The
Wall into an otherworldly
landscape. The even shorter
Window Trail leads to a
natural window overlooking
a deeply cut canyon. And the
more difficult Notch Trail
takes you along the side of a
gully to a break in the rocks,
where you can look out over
White River and the Pine
Ridge Reservation, part of
which lies in the South Unit
of the park, added in 1976.
With few roads and access
only with the permission
of private landowners, the
undeveloped South Unit
is difficult to explore. The
adventurous can, however,
drive onto Sheep Mountain
Table via a dirt road that is
impassible in winter and wet
weather. Stronghold Table
in the South Unit, site of
the last Ghost Dance before
Wounded Knee, is a sacred
place for the Lakota.

Badlands National Park

BANFF

Peyto Lake

Southwestern Alberta, Canada
Established November 28, 1885
1,640,960 acres

Headquarters
Banff National Park
224 Banff Avenue
Banff, Alberta T1L 1B3
Canada
403.762.1550
www.pc.gc.ca/en/pn-np/ab/banff
@BanffNP

Visitor Centers
Banff Visitor Centre (403.762.1550),
in the town of Banff, and Lake Louise Visitor Centre (403.522.3833), in
the town of Lake Louise. Both open
every day except Christmas.

Entrance Fees
Adult, C$10.00, senior citizen
C$8.40, free admission for youth 17
& under; family rate C$20.00.

Accommodations
Thirteen campgrounds offer a
total of over 3,000 campsites. Reservations accepted (877.737.3783);
all others first-come, first-served.
For information on campsite
availability and reservations, visit
national park website. Mosquito
Creek, Rampart Creek, and Silver-
horn Creek Campgrounds have dry
toilets only.

The resort towns of Banff and Lake
Louise, located in the park, offer
a wide range of lodging facilities;
contact Banff & Lake Louise Tourism (www.banfflakelouise.com,
403.762.8421).

When to Go
The park is open all year, but travel
can be restricted in winter due to
road closures. Most people come
in July and August. Some of the
best wildlife viewing is possible
in the fall, when herds of elk and
deer move to winter grounds. Winter sports are popular December
to April in many areas of the park.
Not until early July are the major-
ity of the park's mountain passes
open and dry. Come prepared for
snow regardless of the time of
year—even in August.
And don't forget that Banff's
northern location means that
the length of day varies greatly
throughout the year, ranging
from over 16 hours of daylight
in late June to barely 8 hours in
December.

The first and probably most
famous of Canada's national
parks, Banff is bejeweled
with strings of turquoise-
colored lakes, pristine alpine
meadows, and wild, power-
ful rivers that rush through
forested valleys. From its
famous elk to rarer grizzly
and caribou, wildlife abounds
in these mountains. In the
town of Banff, the unofficial
gateway to the Canadian
Rockies, it's not uncommon
to encounter elk munching
serenely on potted pansies
or cropped hedges. Capping
Banff's range of spectacular,
sheer, glacially tormented
mountains is the largest ice
deposit south of Alaska, the
vast Columbia Icefield – 79
square miles in area and up to
a thousand feet thick.

The park had its begin-
nings in 1883, when three
railway workers stumbled
across a cave and basin
containing hot springs on the
eastern slopes of Alberta's
Rocky Mountains, near the
Bow River. "It's like some
fantastic dream from a
tale of the *Arabian Nights*,"
declared one of the discover-
ers, William McCardell, who
also imagined that by estab-
lishing a bathing resort he
and his partners could bring
attention and tourism to
the area. Today, a cluster
of hotels, restaurants, and
shops have made the old
Canadian Pacific railroad
town of Banff into the seat
of the park, yet the peaks
towering overhead, a bounty
of bike trails, the evergreen
forest, and the swift Bow
River prevent it from seeming

stifling or overcrowded.
Less than an hour's drive
northwest is the park's other
major center, Lake Louise,
which in winter is one of
Canada's top ski resorts. To
see the lake itself, named for
Queen Victoria's youngest
daughter, take Highway 1A
across the river to the usually
jam-packed Chateau Lake
Louise. In summer, rent a
canoe to paddle the blue-
green glacial waters; the lake
rests beneath high, gnarled
peaks and runs up against
the sheer face of Mount Victoria.

Several popular scenic
routes crisscross the park. The
40-mile Bow Valley Parkway
scenic drive, between Banff
and Lake Louise, features a
number of marked trails and
interpretive areas along the
way, giving visitors a good
overview of the park's topography and natural history.
One of the best is the Johnston Canyon Trail, where a
three-and-a-half-mile round-
trip trail, veering out over

swift currents on catwalks,
leads to two spectacular
falls. The Icefields Parkway,
a magnificent 143-mile drive
between Lake Louise and Jasper townsite, has been called
one of the world's greatest
mountain routes and is one
of Canada's highest roads.
With an average elevation
of 5,100 feet, it skirts lakes
and forested river valleys, rising, falling, and penetrating
into the Columbia Icefield. A
good stop is Bow Lake, where
Crowfoot Glacier adorns
the rugged face of Crowfoot
Mountain. Rising to the
chilly 6,785-foot Bow Pass,
the parkway pauses at an
overlook for a stunning view
of Peyto Lake. Three hours
north of Banff townsite, near
Sunwapta Pass, you can walk
(weather permitting) onto
Athabasca Glacier. The entire
Icefields Parkway can be
completed in several hours,
or for a more leisurely tour,
try stretching it into an overnight trip.

SD 101

SD 140

AB 123

AB 142

BIG BEND

© Globe Turner

Southwestern Texas
Established June 12, 1944
801,163 acres

Headquarters
Big Bend National Park
PO Box 129
Big Bend National Park, TX 79834
432.477.2251
www.nps.gov/bibe
@BigBendNPS

Visitor Centers
Panther Junction Visitor Center (432.477.1158), 26 miles from Persimmon Gap entrance, and Chisos Basin Visitor Center (432.477.2264), open all year. Castolon Visitor Center (432.477.2666), Persimmon Gap

Visitor Center (432.477.2393), and Rio Grande Village Visitor Center (432.477.2271), 20 miles south of Panther Junction, open November to April.

Entrance Fees
$30 per vehicle or $25 per motorcycle for 7 days. $15 per person on foot, bicycle, or bus.

Accommodations
Three developed campgrounds. Chisos Basin, Rio Grande Village, and Cottonwood require reservations for all sites (877.444.6777); Cottonwood is closed in summer. RV park operated by concessionaire at Rio Grande Village (432.477.2293) has 25 reservable sites. Chisos

Mountains Lodge (432.477.2291), the only lodging in the park, open all year. For nearby lodging outside the park, contact Alpine Chamber of Commerce (www.alpinetexas.com, 432.837.4144).

When to Go
The park is free of crowds much of the year. Heavy visitation occurs during spring break, generally the second or third week in March; the week between Christmas and New Year's Day can be very busy, as can Thanksgiving weekend. If it rains enough, the desert blooms beautifully in spring.

Sprawling and isolated, Big Bend is a park of boundaries, a land with many faces. In it, desert, river, and mountains come together to create a surprising landscape marked by an unusual diversity of plant and wildlife. Big Bend is full of surprising combinations: Douglas fir and yucca, sandpiper and roadrunner, spiny softshell turtle and javelina, beaver and Mexican long-nosed bat.

The Chihuahuan Desert defines most of the park, with its cactuses, yuccas, and creosote bushes. Rising unexpectedly, the Chisos Mountains harbor pine, juniper, and oak, with temperatures some 20°F cooler than on the desert floor. And creating a verdant horseshoe curve along the park's southern border (also the U. S. – Mexico border) is the mighty Rio Grande – the park's name derives from the river's northward turn here. The river's floodplains and three major canyons – Boquillas, Mariscal, and Santa Elena – provide some of Big Bend's most spectacular scenery, including immense cliffs of eroded red, orange, and brown limestone.

Many species of plants and animals are found in Big Bend at the extremes of their range. Thanks to its place in a flight path from Central and South America, 409 species of birds have been spotted in Big Bend, making it a birders' mecca. Avian stars include the rare Colima warbler, whose only known nesting places are here and in Mexico; the golden-fronted woodpecker; and endangered peregrine

falcons. There are black bears and whitetailed deer in the mountains, javelina and gray fox in the hills, and coyotes and lizards just about everywhere.

The park's remoteness can make visiting problematic, depending on where you're coming from: It's 410 miles west of San Antonio and 325 miles southeast of El Paso. Hiking is really the best way to experience, enjoy, and appreciate Big Bend (there are more than 150 miles of trails). One of the most popular hikes is in Santa Elena Canyon, a massive boxlike gorge on the southwest edge of the park. A moderate one-and-three-quarter-mile trail goes along a rocky ledge into the canyon mouth, rewarding you with striking views of sculpted limestone cliffs.

After your hike, stop at the Santa Elena Canyon Overlook for great views of the 1,500-foot-deep canyon that was worn away by the abrasive silt and gravel carried by the Rio Grande. About six miles east along the road is an overlook onto the adobe buildings of Castolon, a pioneer settlement established by a mixture of Mexican and Mexican-American farmers, American cattlemen, and the U. S. Cavalry. Although no commercial river rafters are based inside the park, float trips down the Rio Grande into Big Bend are available through outfitters at Terlingua and other nearby towns. Multi-day trips include overnight camping on the banks of Boquillas Canyon or Santa Elena Canyon.

Big Bend National Park

BRYCE CANYON

Bryce Canyon National Park

© Globe Turner

South-central Utah
Established September 15, 1928
35,835 acres

Headquarters
Bryce Canyon National Park
PO Box 640201
Bryce Canyon, UT 84764
435.834.5322
www.nps.gov/brca
@BryceCanyonNPS

UT 109
UT 141

Visitor Center
Bryce Canyon National Park Visitor Center (435.834.5322), on main road 1.5 miles inside the park boundary on State Rte. 63, open all year, including major holidays.

Entrance Fees
$35 per vehicle; $30 per motorcycle; $20 per person on foot or bicycle. All fees valid for 7 days.

Accommodations
Two campgrounds, North and Sunset. North is entirely first-come first-served; Sunset requires reservations for all sites (877.444.6777). Sites fill by early afternoon in summer. The Lodge at Bryce Canyon (877.386.4383) is open early March to Dec. 31.

When to Go
Bryce Canyon is home to a number of wildlife species, including mule deer, prairie dogs, and over 200 bird species. May to September are best for viewing wildlife. Small crowds and some beautiful displays of wildflowers make spring and early summer particularly good times to visit. Crowds peak in July and August. The park is at a high elevation, so the weather can be highly variable. Heavy snow can fall from mid-autumn into spring;
lightning is also a year-round danger. During some winters, Alaskan cold fronts descend onto the Colorado Plateau and bring temperatures as low as -20°F.

Two things make Bryce Canyon unusual among parks: Its relatively compact size, and the beautiful fingers of worn limestone, sandstone, and mudstone called hoodoos, which were described by original inhabitants of the land as "red rocks standing like men." Perhaps a better name for the park would be Hoodoo Plateau, as Bryce Canyon is not a canyon at all but a series of amphitheaters carved out of a high land called Paunsaugunt Plateau. The plateau is one among many that extend from the Mesa Verde cliff dwellings in the east to the Sonoran Desert in the south, known collectively as the Colorado Plateau. Ponderosa pines, high-elevation meadows, and spruce and fir forests border the rim of Paunsaugunt Plateau, while panoramic views of three states spread beyond the park's boundaries.

Because of the park's small size, you can drive from one end to the other in less than three hours – except then you'd miss out on some of the 50 miles of trails that explore thousands of colorful spires, fins, pinnacles, and mazes that make this park special. Think of Bryce as basically a day-hike park with a number of connecting trails that allow you to customize the length and duration of your outing. One of the easiest trails is the half-mile section of Rim Trail between Sunrise and Sunset Points. As the name suggests, Sunrise Point is a popular spot

to catch the first rays of the sun as it paints the fluted walls and sculptured spires warm yellows, oranges, and reds. Another favorite place to watch the sunrise is Bryce Point, one of the highest overlooks along the rim of Bryce Amphitheater, the most striking of 12 bowl-shaped canyons where the hoodoo formations cluster in dense stands.

To explore more, combine the Queen's Garden and Navajo Trails for a moderately strenuous two- to three-hour hike that loops into Bryce Amphitheater from Sunset Point to Sunrise Point. This is a good walk for viewing the canyon's scenery and the park's wildflower displays in spring and early summer. In the summer, rangers lead nature walks. If you camp, participate in the night sky programs – the area boasts some of the nation's best air quality. This, coupled with the lack of nearby light sources, creates wonderful opportunities for stargazing. Long-distance views of more than 100

miles can be enjoyed on clear days along the rim of Bryce Canyon. Spring through fall, wranglers lead horseback rides into Bryce Amphitheater along a dedicated horse trail. In winter, the chill is offset by high-altitude sun, bountiful snow, cross-country skiing, and snowshoeing. The Bryce Canyon Winter Festival, three days of indoor and outdoor events held every Presidents' Day weekend, is among the best times to visit the park in winter.

DEATH VALLEY

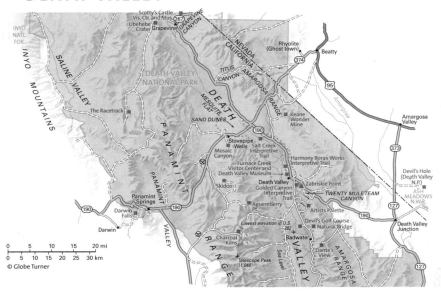

View from Zabriskie Point

Eastern California;
southwestern Nevada
Established October 31, 1994
3,408,396 acres

Headquarters
Death Valley National Park
PO Box 579
Death Valley, CA 92328
760.786.3200
www.nps.gov/deva
@DeathValleyNPS

Visitor Center
Furnace Creek Visitor Center and
Museum (760.786.3200), off Rte. 190
in the center of the park, open daily.
Scotty's Castle Visitor Center is closed
due to flooding.

Entrance Fees
$30 per vehicle for 7 days; $25 per
motorcycle; $15 per pedestrian or
cyclist.

Accommodations
Nine campgrounds. Emigrant, Furnace Creek, Wildrose, and Mesquite
Spring open year-round. Texas
Springs, Sunset, and Stovepipe Wells
open late autumn through spring.
Mahogany Flat and Thorndike open
late spring-autumn. For reservations call 877.444.6777. Other lodging includes Oasis at Death Valley
(800.236.7916), Ranch at Death Valley
(800.236.7916), Stovepipe Wells Village (760.786.7090), and Panamint
Springs Resort (775.482.7680).

When to Go
Not in summer. Even nighttime lows
often top 100°F. The climate is best
November to April, with temperatures between 40 and 75°F. Light to
medium jackets are recommended for
winter. Sunny, dry, and clear weather
predominates, with occasional storms
in winter.

Surely the forbidding name
given to this desert valley
seems ludicrous to anyone
who has ever spent a lovely
spring day here when the sky
was blue, the wildflowers in
bloom, and the temperature
a very reasonable 80°F. But
all it takes is opening your
car window on a sweltering
summer afternoon to appreciate its appropriateness.

The nation's largest
national park south of Alaska,
Death Valley is a land of
extremes. The geological
term for the area is a graben,
a sunken section of the
earth's crust. Its barren, silent
depths contain the lowest
point in North America—
282 feet below sea level, at
Badwater. Also, North America's highest temperature –

134°F – was recorded here;
and with two inches of
rainfall annually, it is the continent's driest spot. Despite
its foreboding name, Death
Valley offers crisp winter air
and lonely, fascinating landscapes, from the sand dunes
near Stovepipe Wells Village
to the mudstone badlands
beyond Zabriskie Point.

The popular belief that
nothing lives in Death Valley
is discounted by the diverse
animal and plant life that has
tenaciously adapted to the
burning heat and dryness.
In fact, over 1,000 species
of plants and trees, including ferns, lilies, and orchids,
flourish inside the park. More
elusive are the many animals,
most of which emerge only
at night. It is not uncommon
to spot bighorn sheep near
Badwater during the cooler
months of the year.

In the heyday of borax
mining in Death Valley, from

1883 to 1889, more than 20
million pounds of the sub-
stance, called "white gold"
by prospectors, were taken
from the Harmony Borax
Works alone. The 20-mule
wagon teams – complete
with seven-foot rear wheels –
that carried borax loads to
a rail depot 165 miles away
are legend. A popular interpretive trail at the Harmony
Borax Works, located near
the Furnace Creek Camp-
ground, leads to the ruins
of the old refinery, offering
a fascinating glimpse into
this colorful era of the valley's history.

Death Valley can be
explored on almost 1,000
miles of paved and dirt roads.
Most of the area's unique
attractions lie no more than
an easy stroll from one of
them. The area is also gain-
ing popularity with cyclists,
drawn to its relatively flat
roads and good weather.

South of the visitor center
you'll find a number of must-
see natural wonders, including the Devil's Golf Course,
which is really the remains of
an ancient lake, and Artists
Palette, a curious terrain of
yellow, red, orange, and even
green rocks that appears to
be painted onto the side of a
canyon. Dante's View, on the
crest of the Black Mountains,
is one of the most spectacular scenic overlooks in the
United States. Scotty's Castle,
at the northern end of the
park, is closed until 2022
due to flooding. Mistakenly
thought to be built by a legendary character known as
Death Valley Scotty, the flam-
boyant, 1922 Spanish-Moor-
ish house cost $2 million and
took ten years to complete.
Horseback riding is available
seasonally out of the Oasis at
Death Valley.

DENALI

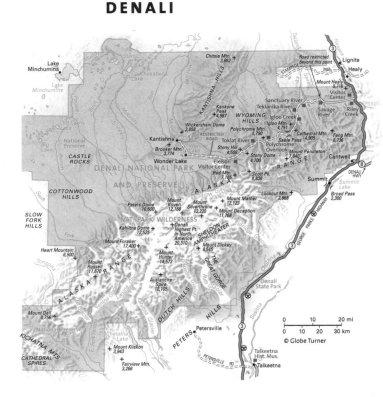

Denali

Central Alaska
Established as Mount McKinley
National Park February 26, 1917
6,075,029 acres

Headquarters
Denali National Park & Preserve
PO Box 9
Denali Park, AK 99755
907.683.9532
www.nps.gov/dena
@DenaliNPS

Visitor Centers
Denali N.P. Visitor Center

(907.683.9532), eastern border
of park, open daily May to mid-
September. Murie Science and
Learning Center (open all year)
serves as the winter visitor center
(907.683.6432). Eielson Visitor
Center, center of park, open June 1
to mid-September.

Entrance Fees
$15 per person for 7 days (age
15 and under free). Varying fees
charged for shuttle and tour bus
services; call 800.622.7275.

Accommodations
Six campgrounds, all open
mid-May or early June to early-
September except Riley Creek,
open all year. Riley Creek, Savage
River, and Teklanika River open to
private vehicles with campground
permits. Sanctuary River, Igloo
Creek, and Wonder Lake accessible
only by shuttle bus. Shuttle seats
and all of the campsites at Riley
Creek, Savage River, Teklanika
River, and Wonder Lake can be
reserved in advance. Denali Park
Hotel (907.683.1700) is open
mid-May to mid-September.
Denali Backcountry Lodge
(800.808.8068) in Kantishna is at
the end of the 92-mile park road.
For other accommodations nearby,
contact the Denali Chamber of
Commerce (www.denalichamber.
com, 907.683.4636).

When to Go
Though the park is open all year,
private vehicle travel is restricted.
By lottery, 300 private cars per day
are allowed to drive the park road
over a 4-day period in September.
The annual road lottery is held in
the fall; applications are accepted
in June (www.recreation.gov). An
extensive shuttle bus system is
available for transportation May-
September. Mosquitoes are worst
in June and early July; after August
15, the atmosphere is crisp and
the annoying bugs are gone. Temperatures average in the mid-60s
during summer, and in winter drop
to -40°F and below. Bring layers of
clothing and rain gear for summer
visits. Winter visits require special
cold-weather gear.

Imagine a park larger than
Massachusetts and you'll
begin to get an idea of how
immense Denali National
Park & Preserve really is.
While it has quickly become
the most popular park in
Alaska, only a few visitors
ever see much more than
a tiny swath of untamed
wilderness on either side of
the 92-mile park road that
runs from the Visitor Access
Center on the eastern border
to Kantishna in Denali's center. Beyond, the vast roadless
and largely trailless wilderness continues on for many
miles in every direction. Biol-
ogist and naturalist Olaus
Murie, overwhelmed by the
expansiveness and range of
natural wonders at Denali,
called it "the greatest scenic
experience on the North
American continent."

Known originally as
Mount McKinley National
Park (the name given to it
by gold prospector William

L. Dickey in 1896), the area
was renamed by Congress in
1980, when it was enlarged
to over three times its original size; the mountain itself
was officially renamed in
2015. Denali, meaning "the
high one," is what early
Athapaskans called the
snowcapped mountain.
Encompassing hundreds
of lakes and rivers, acres of
spruce and birch forests, and
a complete subarctic ecosystem with untold numbers of
large mammals, the park's
greatest spectacle continues
to be mighty Denali (Mt.
McKinley). It rises from a valley only 1,000 feet above sea
level to an altitude of 20,310
feet, making it the tallest
peak in North America.

Unfortunately, due to
near-ceaseless cloud cover,
Denali can be almost as
tough to spot as a wolf – on
average, one in every five
visitors sees a wolf, while one
in every three sees Denali.

August is a popular viewing
month, but don't get your
hopes up.

Because the park restricts
the number of private vehicles allowed beyond the
Savage River Check Station,
at mile 15, the best way to
see Denali is by shuttle or
tour bus. Though it's only 85
miles from the park's eastern
boundary to Wonder Lake,
it's a good 11-hour journey
round-trip. If you take the
shuttle bus you can get off
virtually anywhere along
the route and wait for a
bus going the opposite way.
While wildlife spottings
aren't guaranteed, on a good
day the drive can be like a
visit to an outdoor zoo, with
Dall sheep clinging to moun-
tainsides, moose lingering
in marshy flats, grizzly
bears lumbering by the roadside, and herds of caribou
grazing nearby.

Despite the great size of
Denali, there are few official
trails for hikers. But because
most of the park is open tundra, you can hike just about
anywhere (although some
areas are regularly closed
by park officials to prevent
habitat damage and to
protect wildlife). Some good
areas for day hikes include
Savage River, Primrose Ridge,
Polychrome Pass, and Wonder
Lake. The roughly one-hour
round-trip Horseshoe Lake
Trail, an easy family hike near
the park entrance, winds
gently through a lovely forest of spruce and aspen to
an old oxbow of the Nenana
River.

EVERGLADES

Southern Florida
Established December 6, 1947
1,542,526 acres

Headquarters
Everglades National Park
40001 State Road 9336
Homestead, FL 33034
305.242.7700
www.nps.gov/ever
@EvergladesNPS

Visitor Centers
Four total, three open all year:
the Ernest F. Coe Visitor Center,
12 miles west of Homestead and
Florida City; Shark Valley Visitor
Center, northern end of the park,
35 miles west of the Florida Tpk.
on U.S. 41; and Gulf Coast Visitor
Center, 5 miles south of U.S. 41 on

Hwy. 29. Also, the Flamingo Visitor
Center, open mid-November to
mid-April, 38 miles past the Coe
Visitor Center.

Entrance Fees
$30 per vehicle for 7 days; $25 per
motorcycle; $15 per pedestrian or
cyclist.

Accommodations
No lodging is currently available
in the park. Two developed campgrounds, Flamingo and Long Pine
Key, open year-round, reservations accepted (855.708.2207). For
nearby lodging outside the park,
contact the Naples-Marco Island-
Everglades Convention and Visitors Bureau (www.paradise
coast.com, 800.688.3600) or

South Dade Chamber of Commerce (southdadechamber.org,
305.247.2332).

When to Go
During the dry season, December
to April, the weather is generally clear and pleasant, and
temperatures are moderate. Sun
and insects are bountiful, so bring
sunscreen, protective clothing,
and bug repellent. Mosquitoes can
make summer visits intolerable.

This vast swath of wetlands
and rivers, described as a
"river of grass" by conservationist Marjory Stoneman
Douglas, is the first national
park designated primarily to
protect an ecosystem rather
than for its scenic or historic
value. A wetland of international importance, the
1.5 million acres of watery
wilderness include dense
cypress domes and mangrove
swamps; a saw-grass prairie
that ripples in the wind; and
shadowy hardwood islands
called hammocks, thick with
mahogany and other tropi-
cal trees.

South Florida itself sur-
faced only since the last, or
Pleistocene, ice age, and the
rock beneath the park has

only been exposed for a
mere 6,000 to 8,000 years.
No point in the Everglades is
higher than eight feet above
sea level, and the sole source
of water for the subtropical
region is the rain that falls on
it. The park itself comprises
just one-fifth of the total
mass of the Everglades, and
the rains are being siphoned
away by extensive canal and
levee systems, creating an
artificial drought that threat-
ens the park's wildlife and
ecological balance. Also,
pollutants from agricultural
runoff are contaminating
the plants, animals, and fish
that live in the wetlands. To
fight these trends, Congress,
in one of the world's largest
ecosystem restoration proj-

ects, has extended the park
boundary to help protect
delicate areas.

Today, the Everglades are
home to nearly 40 threat-
ened and endangered spe-
cies, as well as to hundreds
of subtropical plants and
animals found nowhere else
in the United States.
The endangered wood stork
makes its home here – in
decreasing numbers – as
do alligators, snowy egrets,
a few remaining Florida
panthers, and 22 species of
snakes. From May through
August hundreds of female
loggerhead sea turtles come
ashore to lay their eggs on
the Florida beaches. A colorful assemblage of migratory
birds, like numerous warblers,
peregrine falcons, and wading birds, use the park as
a wintering area and a
migration stopover.

The sprawling park has
several distinctive sections:
Everglades City, its western
saltwater gateway; Shark
Valley, which encompasses
the Shark River Slough as
well as the saw-grass prairie;
the area near the historic
town of Flamingo, on the
Florida Bay; and the main
entrance area 12 miles west
of Florida City. Boardwalks
and trails lead off from the
38-mile-long main park road.
Though much of the park
is accessible only by boat
or canoe, a car tour on this
road can work well for short
visits. Also, tram tours from
Shark Valley are a good bet
for wildlife sightings. Boat
tours are available year-
round at Everglades City.

Everglades tidelands

GRAND CANYON

Northern Arizona
Established February 26, 1919
1,201,647 acres

Headquarters
Grand Canyon National Park
PO Box 129
Grand Canyon, AZ 86023
928.638.7888
www.nps.gov/grca
@GrandCanyonNPS

Visitor Centers
On the South Rim, Desert View
Visitor Center (open all year)
and Grand Canyon Visitor Center
(temporarily closed). North Rim
Visitor Center, open mid-May to
late October.

Entrance Fees
$35 per vehicle for 7 days; $30 per
motorcycle; $20 per pedestrian.

Accommodations
Because of heavy crowds during
summer, it is imperative to make
reservations well in advance. On
South Rim, Mather, open all year,
is first-come, first-served December thru February; otherwise
reservations accepted 6 months
in advance. Desert View is open
late April to early October and is
reservation-only. North Rim Campground, open mid-May to October
31; call 877.444.6777 for reservations. Eight lodges in park, including
the historic El Tovar Hotel in the vil-

lage and Phantom Ranch at the bottom of the canyon. Make reservations well in advance; call Xanterra
Parks and Resorts at 888.297.2757.
For information on lodging nearby,
contact Grand Canyon Chamber of
Commerce (grandcanyoncvb.org,
844.638.2901).

When to Go
Facilities on the South Rim are
usually open all year, but conges-
tion is heavy in summer, and air
pollution and hazy days can dimin-
ish visibility considerably. The best
time to visit is spring or fall, when
crowds are slight. For river rafters,
prime season is April through
October.

Grand Canyon National Park

Nothing can prepare you
for standing on the edge of
Yavapai Point on the South
Rim and seeing the Grand
Canyon for the first time –
its vastness is breathtaking.
On a clear day, when visibility
seems unlimited, you'll see
a picture of the earth that
seems to reveal the beginning of time. Stretching 277
miles across northern Arizona, the canyon measures
18 miles at its widest spot,
and is, on average, a mile
deep – making it at once an
unfathomable, shadowy
abyss and a bright panorama
of buttes and sand spires.
At its bottom, some of the
oldest rock in the world sits
exposed, dating back 1.8
billion years. Awed after
his first visit to the Grand
Canyon in 1903, Teddy Roosevelt called it "one of the
great sights which every
American ... should see."

Today, six million visitors
come here annually, though
almost 90 percent never
go further than the two
popular scenic drives on the
canyon's south side. During
peak season, Hermit Road is
closed to private vehicles, but
a shuttle operates at frequent
intervals. The 25-mile Desert
View Drive offers several fine
views of the Colorado River,
including Grandview, Moran,
and Lipan Points. Accessible
only by shuttle, Yaki Point
peers down into the darkly
glowing innermost canyon,
Granite Gorge. The less
visited North Rim is only 10
miles away by air, but 215 by
car. Adventurous travelers
who make it to this side of

the canyon are rewarded
with pristine forests, rolling meadows, and superb,
uncluttered panoramas.
Unfortunately, the North Rim
is open only from mid-May
to late October; its higher
elevation often means heavy
snowfall.

Hiking in the Grand Can-
yon is the reverse of mountain climbing – first you go
down and then you come up.
It's important to remember
this, since it generally takes
twice as much time – and
energy – to come back up
from a hike as to go down.
One of the best family hikes
is the South Rim Nature Trail,
an easy 1.5-mile trek from
the Yavapai Observation Station, set on the brink of the
canyon, to the El Tovar Hotel.
The trail is paved and nearly
level the whole way. One of
the most beautiful hikes is
the North Kaibab Trail, starting in the cool forests of the
North Rim, then descending
through the woods to Roar-
ing Springs. Off Desert View
Drive, Grandview Trail is a
steep but scenic walk to
Horseshoe Mesa, where cop-

per ore was mined around
the turn of the 20th century.
A number of unmaintained
trails in the South Rim's
inner canyons lead to some
beautiful corners of the park,
where solitude and magnificent views await.

An easy way down to the
canyon floor is to go by mule
on one of the popular two-
day trips from the South Rim.
Overnight riders stay and
eat at Phantom Ranch. Trips
can be booked as early as 15
months in advance; call early.
A shorter one-day trip that
goes partway to the river is
also offered. A number of
commercial white-water
trips through the Grand
Canyon begin at Lee's Ferry,
a two-and-a-half-hour drive
from the South Rim, and
range from three days to
three weeks. Some compa-
nies also offer trips starting
from or ending at Phantom
Ranch. River trips operate
April through October.

GRAND TETON

Snake River and Teton Range

Northwestern Wyoming
Established February 26, 1929
310,044 acres

Headquarters
Grand Teton National Park
PO Drawer 170
Moose, WY 83012
307.739.3300
www.nps.gov/grte
@GrandTetonNPS

Visitor Centers
Craig Thomas Discovery &
Visitor Center (307.739.3399), 12

miles north of Jackson on Rte.
89/191/287, open May through
October. Colter Bay Visitor Center and Indian Arts Museum
(307.739.3594), on Jackson Lake,
open early May to early October. Jenny Lake Visitor Center
(307.739.3343), 8 miles north of
Thomas Visitor Center, mid-May
to late September. Flagg Ranch
Information Station (307.543.2372),
2.5 miles south of Yellowstone
National Park's south boundary,
40 miles north of Moose, open
June to September. Laurence

S. Rockefeller Preserve Center
(307.739.3654), 4 miles south of
Moose, open June to September.

Entrance Fees
$35 per vehicle for 7 days; $30 per
motorcycle; $20 per pedestrian.

Accommodations
Six campgrounds. All open from
May or early June to September
or October. All are first-come,
first-served. Grand Teton Lodge
Co. (307.543.3100) runs 4 lodges
in park. For other accommodations, call Signal Mountain Lodge
(307.543.2831) or Triangle X Ranch
(307.733.2183).

When to Go
The high season runs July to Labor
Day, when day temperatures are in
the 70s and 80s, nights in the 40s,
though fishing is best in September. The first heavy snow falls by
Nov and continues through March.
Snow and frost are possible any
time of year, so bring rain gear.

name from lonely 19th-
century French-Canadian
trappers who imagined a
likeness to breasts in the
snowy peaks: Teton is an
archaic French word for
cow's teats. At the feet of the
range sits a string of pleasant,
stream-fed lakes, beyond
which lies Jackson Hole. An
excellent way to see the lay
of the land and appreciate
the area's unusual geology
is to visit Signal Mountain,
accessed by a turnoff near
the south end of Jackson
Lake. This low mountain
rises alone, 800 feet above
the valley; and from its summit, a broad panoramic view
reveals the Snake River curving gently down the length
of the valley floor as well as
the abrupt transition from
flat to mountain range. From
here, it is easy to understand
why early trappers thought
of it as a "hole." If you're on
a tight schedule, the Signal
Mountain Road is a good bet.
In the 50-mile-long valley,
sagebrush flats and forests
of lodgepole pine and spruce
make ideal habitats for
pronghorn, deer, elk, and even
moose. Where the Snake River
braids, wetlands support
trumpeter swans, sandhill
cranes, Canada geese, and a
plethora of ducks.
More than 200 miles
of marked trails offer an
unlimited choice of hiking
or cross-country skiing excursions, depending on the
season. A half-day excursion
runs from the Moose Visitor
Center to Jenny Lake Scenic
Drive, which many consider
to be the visual heart of the
Tetons, offering majestic
views of the central peaks.

Cathedral Group Turnout is
a good lookout. Jenny Lake
is also an excellent spot for
a hike. A good family walk
takes you to outstanding
views of the glacially carved
lake. A longer, six-mile hike
joins the Cascade Canyon
Trail to the spectacular
Hidden Falls and Inspiration Point. A passenger boat
departs from the southern
end of the lake and gets you
to within a half mile of the
falls.
The Jenny Lake Ranger
Station is the center for
mountaineering in the
area. Mountain guides offer
training that has beginners
crawling up sheer cliffs their
first day. Special instructions are required to go on
the guided overnight climbs
of Mount Owen, Mount
Moran, or Grand Teton. Parkway concessionaires and
operators provide a variety
of floating and fishing trips
on the Snake River. Half-day
trips from Pacific Creek usually include a lunch at Deadman's Bar. Horseback rides
are offered out of Colter Bay,
Jackson Lake Lodge, Jenny
Lake Lodge, and Headwaters
Lodge. Winter, though harsh,
is a good time for sports
like cross-country skiing,
snowshoeing, dogsledding,
and snowmobiling and for
spotting large herds of elk
moving across the open flats
to their wintering area south
of the park.

The Tetons take their

WY 120
WY 142

GREAT SMOKY MOUNTAINS

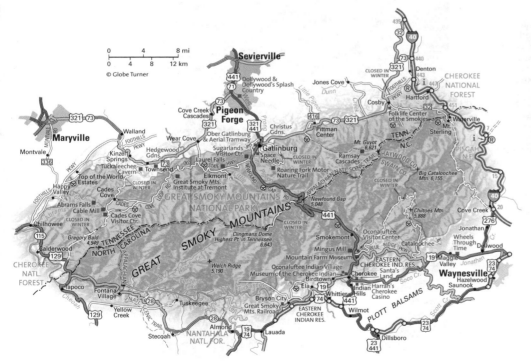

Great Smoky Mountains National Park

**Western North Carolina;
eastern Tennessee**
Established June 15, 1934
522,427 acres

Headquarters
Great Smoky Mountains
National Park
107 Park Headquarters Road
Gatlinburg, TN 37738
865.436.1200
www.nps.gov/grsm
@GreatSmokyNPS

Visitor Centers
Oconaluftee Visitor Center
(828.497.1904), 2 miles north
of main park entrance near
Cherokee, NC. Cades Cove
Visitor Center, 12 miles south of
Townsend, TN. Sugarlands
Visitor Center (865.436.1291),
2 miles south of Gatlinburg, TN,
on Newfound Gap Rd. All visitor centers open daily except on
Christmas Day.

Entrance Fees
None.

Accommodations
Ten campgrounds, with a total
of 939 campsites. Cades Cove
and Smokemont open all year;
others generally open from
Spring to October. All require
reservations at least April/May-
Oct (877.444.6777), year-round at
Cades Cove and Smokemont.

Length of stay limited to 14 days.
Advance reservations required for
LeConte Lodge (865.429.5704),
accessible only by foot; open
mid-March to mid-November. For
nearby lodging outside the park,
contact Gatlinburg Convention
and Visitors Bureau (www.gatlin
burg.com, 800.588.1817).

When to Go
Although summer is the most
popular time of year, spring and
fall are special seasons to visit.
Summer can be hot, humid, and
crowded. Spring can be wet, but
is an ideal time for blooming wildflowers. Fall, when the hardwood
forests put on their magnificent
foliage display, is often very
crowded, but temperatures are
much cooler. Park roads close in
icy conditions. Note that park
elevations range from 875 to 6,643
feet, so the local topography can
dramatically affect the weather,
with temperature variations of
10–20 degrees possible from
mountain base to top.

Named for the distinctive
smokelike haze that envelops them, the Great Smoky
Mountains are the stately
ceiling of the Appalachian
Highlands, encompassing
16 peaks that rise more than
6,000 feet above sea level.
The mountains' dense nests
of brush and trees, packed
with leaves that exude water,
oxygen, and hydrocarbons,
are responsible for the
"smoky" air that is especially
visible after rain or in the
early morning. Its persistent
blue haze only adds to the
mystique of the Smokies, a
jumble of ancient mountains
whose pioneer history and
lofty isolation draws nine
million visitors each year –
the most of any of the country's national parks.
The park straddles North
Carolina and Tennessee and is
one of the largest protected
land areas east of the Rocky
Mountains. In its central
section, the transmountain
Newfound Gap Road
(U. S. 441) links the Sugarlands and Oconaluftee Visi-

tor Centers, climbing from
2,000 feet to 5,048 feet, and
provides spectacular views of
the surrounding peaks. Just
past the Oconaluftee Visitor
Center, in the southern section of the park, is Mountain
Farm Museum, a collection
of farm buildings where costumed interpreters reenact
pioneer farm life (summer
through late October). At
nearby Mingus Mill, a costumed miller demonstrates
how pioneers ground grain.
In the western section, at
Cades Cove, an isolated
valley first settled in 1819,
rangers maintain a historical
and cultural preserve of log
cabins, churches, and other
buildings.
With 850 miles of trails
in the Smokies, ranging
from short paths to a
71-mile-long segment of the
Appalachian Trail, getting
out of the car and walking is
a great way to dip into what
is referred to as a "little bit
of the world as it once was."
Among the easiest hikes are
several gentle, quarter-mile
paths called Quiet Walkways.
An eight-mile round-trip hike
along the Appalachian Trail
between Newfound Gap
and Charlies Bunion passes
through a spruce forest and
provides mountain vistas.
The paved, two-and-a-half-
mile round-trip Laurel Falls
Trail, the most popular waterfall trail in the park, meanders
through stands of pine and
oak, while the strenuous,
eight-mile round-trip Ramsay
Cascades Trail leads to the
park's highest waterfall.

NC 82
TN 103
NC 139
TN 140

HOT SPRINGS

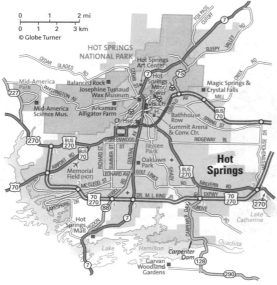

0   1   2 mi
0   1   2   3 km
© Globe Turner

Bathhouse Row, Hot Springs

Western Arkansas
Established March 4, 1921
5,554 acres

Headquarters
Hot Springs National Park
101 Reserve St.
Hot Springs, AR 71901
501.620.6715
www.nps.gov/hosp

Visitor Center
Fordyce Bathhouse Visitor Center
(501-620-6715), in the middle of
Bathhouse Row, open daily including major holidays.

Entrance Fees
None, but donations accepted.
Concession fees charged for the
thermal baths.

Accommodations
Gulpha Gorge campground,
open all year, first-come, first-served. Hot-springs water is piped
into Arlington Resort Hotel &
Spa (800.643.1502); Hotel Hale
(501.760.9010); Hotel Hot Springs
(877.623.6697). For other nearby
baths/spas and lodging, contact
the Hot Springs Convention and
Visitors Bureau (www.hotsprings.
org, 800.772.2489).

When to Go
Spring is mild and begins in Febru-
ary, when four-petaled bluets, the
first of many wildflowers, begin
to bloom. Summers are hot and
very humid, and July is the most
crowded month. Try going in late
fall, when mountains in the area
produce spectacular foliage.

Hernando de Soto has often
been credited with discovering the springs in 1541, but
Native American tribes may
have taken advantage of the
curative powers of the hot
flowing water long before he
came to the area. The 47 hot
springs that flow from the
southwestern slope of Hot
Springs Mountain at a temperature of 143°F came to
the attention of President
Andrew Jackson in 1832,
which is when he set aside
the area as a "special reservation." Some think this
makes the park the oldest
in the National Park System,
predating Yellowstone's
establishment by 40 years.
  Hot Springs' main attraction has always been its magnificent bathhouses, many of
them built in the 1930s and
'40s, when a million people
per year would come to soak
in the mineral-rich waters.
Eight historic bathhouses
have been preserved along
a stretch of Central Avenue
dubbed Bathhouse Row, but
only one – the Buckstaff –
still offers the traditional
treatment, including a hot
pack, steam cabinet, and
needle shower. Even if you
don't feel like a soak, stop by
the Fordyce Bathhouse, which
now serves as the park's
visitor center. This restored
"temple of health and beauty" contains three floors of
fountains, stained-glass windows, authentic furnishings,
and elegant statuary.
  But the hot springs aren't
the park's only attraction.
There are 26 miles of day-use hiking trails through
beautiful woodlands – all
the more remarkable when
one considers that the park
is basically in the center of a
busy city (former home of Bill
Clinton, the 42nd president
of the United States). From
De Soto Rock, near the corner
of Central Avenue and Fountain Street, follow the trail to
the Hot Water Cascade. The
flowing water here began
its journey as long as 4,000
years ago, when rainfall
seeped through fractures
in the earth. The tufa rocks
you see here, a result of the
water's mineral content, are
building up at the rate of
about one-eighth inch a year.
Nearby, running behind
the bathhouses along a
wooded hillside, the Grand
Promenade is a landscaped,
brick walkway that leads
to walking trails, concealed
springs, and pleasant views
of the downtown skyline.

MAMMOTH CAVE

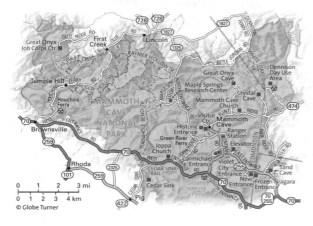

0   1   2   3 mi
0 1 2 3 4 km
© Globe Turner

Mammoth Cave National Park

South-central Kentucky
Established July 1, 1941
54,012 acres

Headquarters
Mammoth Cave National Park
P.O. Box 7
1 Mammoth Cave Parkway
Mammoth Cave, KY 42259
270.758.2180
www.nps.gov/maca
@MammothCaveNP

Visitor Center
Mammoth Cave Visitor Center
(270.758.2180), center of park, open
daily all year except
Christmas.

Entrance Fees
None. A variety of fees ($4–$60)
are charged for cave tours.
Advance reservations strongly recommended; visit www.recreation.
gov or call 877.444.6777.

Accommodations
Reservations available for Maple
Springs (groups only, accommodates horses) and Mammoth
Cave, May through November;
and Houchin Ferry, year-round; call
877.444.6777. For reservations at
the Lodge at Mammoth Cave, call
844.760.2283. For nearby lodging
outside park, contact the Cave City
Tourist and Convention Commission (cavecity.com, 270.773.8833).

When to Go
The park is least crowded before
Memorial Day and after Labor Day,
although these times can be cool
and wet. Mid-March to mid-April,
when the redbud and dogwood
trees are in bloom, is a popular
time to visit. Summers tend to be
hot, humid, and buggy. The temperature in the cave varies somewhat, but hovers around 54°F.

The world's longest known
cave system, Mammoth has
over 412 mapped miles of
underground passages. The
14 miles of cave open to the
public offer a rare glimpse
of a beautiful subterranean
world, including stalagmites
and stalactites, canyons, and
mineral-encrusted chambers.
  Above the cave is a rugged
country of second-growth
forest and riverways providing sanctuary to about 1,300
plant species. The Green
River flows for 27 miles
through the park and has
played an important part
in Mammoth's formation.
Until about 280 million years
ago, a shallow sea covered
the area, and left in its wake
deep limestone beds, the
remnants of mucky ooze
that once covered the sea
floor. As the sea receded,
a layer of sandstone was
deposited above the limestone. Rainwater made
acidic by deposits of carbon
dioxide in the soil drained
to the ever-lowering Green
River watershed and slowly
dissolved the limestone,
creating the caves. At depths
of up to 360 feet below the
surface, cave streams are still
forming passages today. The
sandstone forms a protective
roof overhead and in effect
makes the area a cave rather
than a pit.
  As the caves spread,
numerous aquatic species
began to adapt to the underground kingdom. Of approximately 115 species that use
the cave on a regular basis,
many of them are troglobites (strictly cave-dwelling),
and a few are found only in
Mammoth Cave and its
immediate vicinity. Three
endangered species – the
Kentucky cave shrimp, the
Indiana bat, and the gray
bat – also make their homes
here.
  Native Americans discovered the cave system approximately 4,000 years ago and
mined minerals, including
gypsum and selenite, from
it over the next 2,000 years.
Settlers rediscovered the
cave in the late 1790s and
during the War of 1812, slaves
mined saltpeter to use in the
manufacture of gunpowder.
Local entrepreneurs began
giving tours in 1816. Now the
park offers a variety of interpretive cave tours of varying
lengths and difficulty. Among
the more popular are the
two-hour, two-mile Historic
Tour, focusing on the area's
human history; the one
and one-quarter-hour, one-quarter-mile Frozen Niagara
Tour, which explores the
pits, domes, and decorative
dripstone formations of the
Frozen Niagara section of
Mammoth Cave; and the
River Styx Tour, two-and-half
hours focusing on the unique
geology and natural history
of the cave. Surface activities
include 60 miles of trails to
explore on the north side of
the park.

MESA VERDE

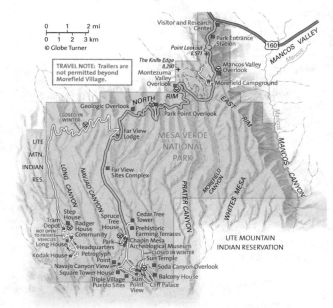

TRAVEL NOTE: Trailers are not permitted beyond Morefield Village.

Mesa Verde, Spanish for "green table," was the first cultural park set aside by the United States. It's a stirring experience to walk through ancient dwellings and numerous mesa-top villages built by Ancestral Puebloan peoples between A.D. 600 and 1300. Though crumbling villages were first recorded by a U.S. Army lieutenant in 1849, and scientific examinations of the sites occurred in 1874, it wasn't until two local cowboys tracked stray cattle after a December 1888 snowfall that the most magnificent and largest of the structures – Cliff Palace – was discovered. Though many of the inhabitant's secrets have been learned since, the answer to the biggest mystery, why the Mesa Verde people left their ancestral home, may never be fully understood.

A short trip to Mesa Verde should include a stop at the Chapin Mesa Archeological Museum, located at the Spruce Tree House Trailhead, followed by a drive over the Mesa Top Loop Road. The museum contains excellent dioramas that help bring the Mesa Verde people to life. There's also a good collection of Mesa Verde pottery, decorated in signature black geometric designs against a white background. The 12-mile Mesa Top Loop Road winds through Chapin Mesa's fragrant piñon-juniper woodland. This area contains the largest concentration of sites. Plan to visit Cliff Palace, the largest cliff dwelling in North America, which once housed more than 100

Mesa Verde National Park

Ancestral Puebloan peoples. Because of ladders and staircases at Cliff Palace, it is not recommended for visitors with disabilities. Other highlights on the loop include the Square Tower House overlook: A short trail leads to a dramatic viewpoint above the park's tallest tower, the four-story remnant of an even larger, multitiered structure. Another good stop is the Sun Temple, a strange, doorless structure that was never inhabited, but may have been a ceremonial center.

Because of the fragile nature of the sites and the emphasis on the cultural

aspects of Mesa Verde, hiking is somewhat restricted in the area. But if you're up for it, take the 2.4-mile Petroglyph Point Trail, a self-guided nature walk that leads from the museum around the base of the cliff on the east side of Spruce Tree House and Navajo Canyon. Register at the museum before you go. Before leaving Mesa Verde, be sure to stop at 8,572-foothigh Park Point fire lookout. The highest spot in the park, it affords open views into Arizona, Utah, New Mexico, and Colorado, which all meet at nearby Four Corners.

Southwestern Colorado
Established June 29, 1906
52,485 acres

Headquarters
Mesa Verde National Park
PO Box 8
Mesa Verde, CO 81330
970.529.4465
www.nps.gov/meve

Visitor Centers
Mesa Verde Visitor and Research Center (970.529.4465), at the entrance to the park, open daily June to December. Chapin Mesa Archeological Museum, at southern end of park, closed temporarily; normally open daily all year, except major holidays.

Entrance Fees
Fees change seasonally. Go to www.nps.gov/meve/planyourvisit/fees.htm. Tickets for tours of Cliff Palace, Balcony House, and Long House are available online at recreation.gov ($7).

Accommodations
Morefield Campground, open mid-April thru mid-October; reservations accepted (800.449.2288). Lodging available at Far View Lodge, closed October–April. Call 800.449.2288.

When to Go
With wildflowers in bloom, April through September is best. Many facilities and cliff dwellings are closed in winter, though crosscountry skiers have access to park trails, conditions permitting.

OLYMPIC

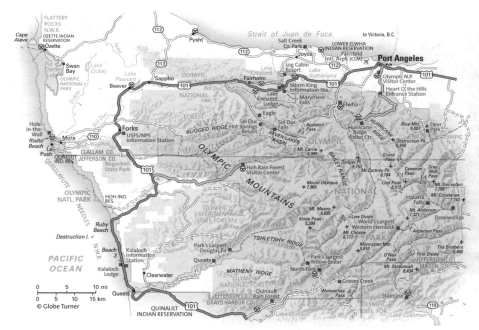

Trail through Hoh Rain Forest

marine fossils are embedded in the summits. The mountains so effectively trap marine moisture that the western slopes boast the wettest climate in the lower 48 states.

To see this often overwhelming park, plan to spend at least two days. On clear days, the subalpine meadows at Hurricane Ridge offer tremendous views of peaks and glaciers in the distance. Watch for blacktailed deer and for Olympic marmots, which whistle when approached (one of several animals unique to the Olympics). Boating and trout fishing are popular at lovely Lake Crescent, carved by a glacier (although Indian legend says that the angered Mount Storm King created it when he hurled a boulder at two fighting tribes).

The valleys of the Queets, Quinault, and Hoh Rivers contain some spectacular examples of Sitka spruce as well as Douglas fir, western red cedar, big-leaf maple, red alder, vine maple, and black cottonwood. In the Hoh Rain Forest, trees enrobed in club moss and licorice ferns stretch 300 feet to the sky.

More than 70 miles of Pacific Ocean coastline also invite exploration: wild shoreline, thunderous surf, arches and sea stacks, mesmerizing tidal pools, and the calls of gulls, bald eagles, and black oystercatchers. The trails to Ruby or Rialto Beaches lead to fine examples of wild, windswept shores.

A dizzying and diverse labyrinth of soaring peaks, moss-covered forests, lush meadows, alpine lakes, and fog-shrouded coastline, Olympic National Park encompasses three ecosystems and almost a million acres. Occupying the central portion of the Olympic Peninsula, as well as a narrow 73-mile strip of land along the Pacific coastline, the park lies near the northwesternmost tip of the lower 48 – a silent, thriving, wondrous place that can be as eerie as it is awesome.

Before the first Europeans set foot here – probably Juan de Fuca, in 1592, though George Vancouver made the first intensive investigation

of the waterways 200 years later – Native Americans spent thousands of years in what is now the park. Today, just as when native peoples lived here, it remains a place of solitude and discovery. These characteristics are preserved in part due to its immense size, but also because of its isolation – still crowned by alpine glaciers and surrounded on three sides by water. Indeed, this is a place intrinsically tied to the water. Glaciers gouged out Puget Sound and Hood Canal to the east and the Strait of Juan de Fuca to the north, isolating the peninsula from the mainland. The rock of the Olympic Mountains developed under the ocean;

Western Washington
Established June 29, 1938
922,649 acres

Headquarters
Olympic National Park
600 East Park Avenue
Port Angeles, WA 98362
360.565.3130
www.nps.gov/olym
@OlympicNP

Visitor Centers
Olympic National Park Visitor Center (360.565.3130), in Port Angeles, near Race St. off U.S. 101, open and staffed all year. Hurricane Ridge Visitor Center, 17 miles past visitor center in Port Angeles, and Hoh Rain Forest Visitor Center (360.374.6925), 19 miles east of

U.S. 101 on the west side of park, open daily in summer, intermittently during fall-spring, road and weather conditions permitting.

Entrance Fees
$30 per vehicle; $25 per motorcycle; $15 per person on foot or bicycle. All fees valid for 7 days.

Accommodations
Fourteen campgrounds, with a total of 841 sites, most available first-come, first-served; group reservations accepted for Hoh, Kalaloch, and Sol Duc (summer only). Some campgrounds operate seasonally; visit recreation.gov for information on campground availability. Four lodges and resorts also operate inside the park. For

information on these and other nearby lodging outside the park, contact Olympic Peninsula Tourism Commission (olympicpeninsula.org, 800.942.4042).

When to Go
Olympic National Park is open 24 hours a day, 365 days a year. December and January are the quietest months, but some roads may be subject to winter closure due to snow; for visitors, proper weather gear is a must. Late spring, when colorful subalpine meadows are beginning to bloom and snow still bedecks the craggy mountain peaks, is an ideal time for a visit. Summer weekends have heavy traffic; call ahead for road conditions, even in April and May.

REDWOOD

Northwestern California
Established October 2, 1968
138,999 acres (includes 3
state parks)

Headquarters
Redwood National and State Parks
1111 Second Street
Crescent City, CA 95531
707.464.6101
www.nps.gov/redw
@RedwoodNPS

National Park Visitor Centers
Crescent City Information Ctr. (707.465.7335), at 111 Second St., Crescent City; Thomas H. Kuchel Visitor Ctr. (707.464.6101), at southern end of park; Hiouchi Visitor Ctr. (707.458.3294), Hwy. 199 in Hiouchi; and Prairie Creek Visitor Ctr. (707.488.2039), off Hwy. 101 north of Berry Glenn, open all year except major holidays. Jedediah Smith Visitor Ctr., Hiouchi, open daily in summer.

Entrance Fees
None for national park. $10 day-use fee in developed areas at state parks inside border.

Accommodations
Four state-run campgrounds are inside the park. All open all year except Mill Creek, open mid-May through September. Reservations required (www.reservecalifornia.com, 800.444.7275). For area lodging, contact chambers of commerce listed on park website.

When to Go
Redwood draws big crowds June through September, so the best bet is to visit in fall, when the deciduous trees change color. In winter, the park is cool and often drenched with rain. Rhododendrons flourish during the spring, and migrating birds bring the big trees to life in both spring and fall.

It's hard to imagine that anyone ever said if you've seen one tree, you've seen 'em all – especially once you have visited the glorious old-growth groves of Redwood National and State Parks. Called "ambassadors from another time" by John Steinbeck, redwoods are almost immortal. Their foot-thick bark repels fire and insects; they cannot be killed by disease; when one falls, new sprouts regenerate from its stump and trunk. In the park, acres of the earth's tallest living things cluster around you, bearing silent witness to thousands of years of both human and natural history.

The hundreds of groves stretching along the northernmost reaches of the California coast are remnants of a magnificent primeval forest that once covered two million acres. Today, only a fraction of that remains. In 1978, Congress expanded the park by an additional 48,000 acres, but most of the new land had already been logged, leaving one park official to describe it as having the "look of an active war zone." Since then, many acres have been replanted and logging roads eliminated, but it will take at least another two or three centuries for the slow-growing seedlings to reach even modest size. Three of the tallest trees in the world, led by Hyperion at 381 feet, are located in the park, but, fearing ecosystem damage or vandalism resulting from human traffic, none of the exact locations have been made public, but instead are closely-guarded secrets.

The aptly named Tall Trees Grove is accessible by car, via a 16-mile drive that takes about 50 minutes (50 free permits are given out per day for car access to the grove; permits are issued via the park website). You'll want to walk down the trailhead to Redwood Creek for some close-up time with these giants.

More than 200 miles of hiking trails in the park include several shorter, looped nature trails with exhibits, all near roadside pullovers. One of the best is a mile-long walk through the Lady Bird Johnson Grove, leading to the 1968 National Park dedication site. Kids enjoy the cool, moist air around the trees and are

fascinated by the hollowed-out redwoods that continue to live and grow. These living caves once sheltered the fowl and livestock of early settlers. For a more active look at the area, join one of the ranger guided walks offered June through August. Fresh-water and ocean fishing and kayak rentals are also available in the summer.

A drive through the towering redwoods.

ROCKY MOUNTAIN

Rocky Mountain National Park

Northern Colorado
Established January 26, 1915
265,807 acres

Headquarters
Rocky Mountain National Park
1000 Highway 36
Estes Park, CO 80517
970.586.1206
www.nps.gov/romo
@RockyNPS

Visitor Centers
Beaver Meadows Visitor Center, U.S. 36 at east entrance, open daily. Kawuneeche Visitor Center, U.S. 34 north of Grand Lake, open daily summer. Fall River Visitor Center, U.S. 34 near east edge of park; Moraine Park Discovery Center, on Bear Lake Rd. south of Beaver Meadows entrance; and Alpine Visitor Center, on U.S. 34, open late spring to mid-autumn.

Entrance Fees
$35 per vehicle for 7 days; $30 per motorcycle; $20 per pedestrian or cyclist.

Accommodations
Longs Peak and Timber Creek first-come, first-served. Reservations necessary in summer at Aspenglen, Glacier Basin, and Moraine Park; call 877.444.6777. Only Moraine Park is open all year; others, May to September or October. For nearby lodging outside the park, call Estes Park Visitor Center (visitestespark.com, 800.443.7837) or Grand Lake Chamber of Commerce (gograndlake.com, 800.531.1019).

When to Go
Half of the 4.7 million annual visitors come mid-June to mid-August. September may be the best time to visit; elk move to lower elevations, where visitors can hear the bull elks' mating bugles.

Congress had two goals in mind when it established Rocky Mountain National Park in 1915. First, to protect the rugged environment and dwindling stocks of wildlife, and second, to promote the recreational opportunities of a land so high that the tallest summits are perpetually capped in snow and even the valleys are 8,000 feet above sea level. For hundreds of years the land's high altitude and extreme weather made it off limits to all but the boldest of travelers. With the construction of mountain roads and a grand hotel near Estes Park around the turn of the 20th century, unrestricted development seemed inevitable – until the fiery pioneer naturalist Enos

Mills enlisted the aid of John Muir to lobby for the creation of a new park.

Straddling the Continental Divide, the park contains the source of the Colorado River, which flows south and west into the Gulf of California, and the headwaters of the Cache la Poudre and Big Thompson Rivers sit in the alpine peaks on the east side. Within the park's 415 square miles are 77 named peaks higher than 12,000 feet; 20 of them over 13,000 feet. The tallest, Longs Peak, rises 14,259 feet, and served as a guidepost to pioneers. With 147 lakes to enjoy, and more than 350 miles of trails, the park is an outdoor-lover's dream, particularly for those willing to take on the backcountry, where you're likely to encounter more mule deer than humans.

Even if you never leave your car and stick only to the Trail Ridge Road, it would be worth the visit. Trail Ridge Road follows an ancient path where Native Americans once hunted elk, deer, and beaver. This road ascends to an elevation of 12,183 feet, making it the highest continuous paved road in the U. S. Considered one of the great scenic roads in the country, it was named one of the first All-American Roads in 1996. Depending on snowfall, it usually is open from Memorial Day into October.

For 11 of the road's 48 miles you travel above timberline, winding through the alpine tundra of delicate grasses and wildflowers.

Many of these plants are identical to ones found in the Arctic. A short nature trail, Tundra Communities, begins at the parking area at Rock Cut, where you can peer into a glacier-carved valley 2,500 feet below.

The park offers trails suited to every hiking ability, but because of the high elevations, be alert to altitude sickness and don't attempt any strenuous hikes before acclimating yourself for at least a day. Families might consider the easy Sprague Lake hike, located off Bear Lake Road. It's a half-mile nature walk with spectacular views of Continental Divide peaks. Notably, the park's highest mountain, Longs Peak, is a popular hike, fully accessible to those in condition without technical climbing gear once the ice melts – usually by mid-July. Another good day hike is the four-mile roundtrip Gem Lake Trail, on the park's east side, offering views across Estes Valley to Longs Peak. The lake itself is a fine picnic spot.

There's also excellent rock climbing in the park. One of the most popular areas is Lumpy Ridge, a subalpine outcrop of sheer rock faces two miles north of Estes Park. The Holzwarth Historic Site on the west side of the park offers visitors a glimpse into the life of high-elevation Colorado homesteaders, with daily tours and interpretive programs from mid-June to Labor Day.

SHENANDOAH

Northern Virginia
Established December 26, 1935
199,224 acres

Headquarters
Shenandoah National Park
3655 U. S. Highway 211E
Luray, VA 22835
540.999.3500
www.nps.gov/shen
@ShenandoahNPS

Visitor Centers
Dickey Ridge Visitor Center, at Mile 4.6; Harry F. Byrd, Sr. Visitor Center, at Mile 51. Mileposts, numbered from north to south on Skyline Drive, help visitors locate park facilities, services, and areas of interest.

Entrance Fees
$30 per vehicle for 7 days; $25 per motorcycle; $15 per pedestrian or cyclist.

Accommodations
Five campgrounds, open late March to early May thru Oct./Nov. Matthews Arm, Loft Mountain, and Dundo have reservable sites; Big Meadows and Lewis Mountain are first-come, first served only. Two lodges and one cabin complex; reservations essential. Contact DNC Parks & Resorts at Shenandoah (877.847.1919).

When to Go
Portions of Skyline Dr. may close temporarily after snowfall. Wildflowers abound by the roadside seasonally. Fall foliage is spectacular and draws large crowds, particularly in October. To avoid traffic, it's best to come early on a weekday.

Early autumn at Shenandoah

Shenandoah National Park sits in the northern part of Virginia's Blue Ridge, a flank of the Appalachian Mountains that was inhabited by farmers for more than a century. Skyline Drive, a two-lane road that rides the crest of the Blue Ridge Mountains for 105 miles through the length of the park, offers views of the Shenandoah River and Massanutten Mountain to the west, and the rolling Virginia Piedmont country to the east. Drivers enjoy contrasting up-close views of forest with wide-open vistas. The Appalachian Trail, a 2,190-mile-long footpath that stretches from Maine to Georgia, roughly parallels about 101 miles of Skyline Drive. For hikers on the trail or motorists wending their way along the drive, Shenandoah offers the sense of traveling high above the world in a separate and serene environment – especially during the dazzling mountain laurel and azalea displays of late spring and early summer.

Human habitation can be traced back approximately 11,000 years. The first European settlers arrived soon after Governor Alexander Spotswood of Virginia led an expedition of 63 men across the Blue Ridge in 1716. By 1800, the lowlands had been settled by farmers who spread into the mountains as valley farmland became scarce. By the 20th century, those who had cleared the land began to leave as both game animals and the soil thinned. In dedicating the park in 1936, President Franklin D. Roosevelt committed the federal government to something new in land management – returning to forest a huge tract of acreage that had been used for farming, grazing, and timbering.

The experiment was successful. Today, there are some 1,400 plant species in Shenandoah. More than 270 species grow at Big Meadows (Mile 51) alone, the largest treeless area in the park, an excellent place to spot wildlife, pick blueberries and huckleberries, and see flowers and plants such as deerberry bushes and panicled dogwood. Rocks that exhibit columnar jointing, a phenomenon created by the rapid cooling of molten lava, can be seen at numerous points, including Compton Peak (Mile 10.4) and Franklin Cliffs (Mile 49). There are a number of spectacular waterfalls: 93-foot Overall Run (Mile 22.2) is the park's highest. There are over 500 miles of trails in the park. From Mile 52.5, hike the two-mile Mill Prong Trail to Rapidan Camp, used by President Herbert Hoover as a retreat until he donated it to the park.

VOYAGEURS

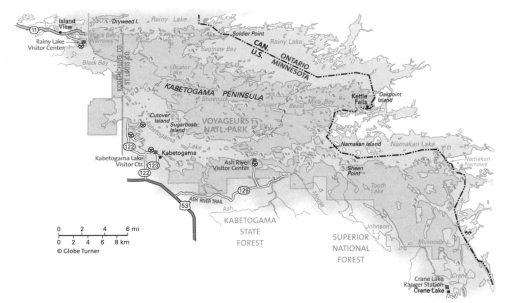

Northern Minnesota
Established April 8, 1975
218,222 acres

Headquarters
Voyageurs National Park
360 Highway 11 East
International Falls, MN 56649
218.283.6600
www.nps.gov/voya

Visitor Centers
Rainy Lake Visitor Center, east of International Falls off Hwy. 11, open all year with limited winter hours (218.286.5258).; Kabetogama Lake Visitor Center (218.875.2111), off U.S. Rte. 53 on County Rd. 123/Gappa Rd. and Ash River Visitor Ctr. (218.374.3221), off Ash River Trail, open May to Sept.

Entrance Fees
None. A Minnesota fishing license is required for fishing inside the park (www.dnr.state.mn.us/licenses/fishing).

Accommodations
Kettle Falls Hotel (218.240.1724) is accessible by boat, floatplane, snowmobile, or skiplane only. The park has 147 frontcountry and 14 backcountry campsites, all reservable (recreation.gov, 877.444.6777). Houseboats can be rented, and require a park permit with $10 per night fee. For nearby lodging outside the park, contact International Falls Area Chamber of Commerce (www.ifallschamber.com, 800.325.5766).

When to Go
The waterways begin to open in late April. Migratory birds return to the park in summer, and fall is an ideal time to enjoy the foliage. The fishing is best late May to June and September to October; boating is great spring, summer, and fall. Summer storms can arise swiftly, creating high waves and making travel hazardous on the large lakes. In winter, beware of slush and thin ice; stay on marked and staked trails. Use caution in late autumn and early spring when ice conditions can be unstable.

One of the country's least visited national parks is this 55-mile-long swath of boreal forest, glacier-carved lakes, and pine-covered islands that straddles the Minnesota–Ontario border. Named for the 18th- and early-19th-century French-Canadian fur traders who canoed and trapped in the region, it is the only part of the National Park System wholly within the Arctic watershed of Hudson Bay. The labyrinth of waterways and marshes consists primarily of four large lakes and the park's centerpiece, Kabetogama Peninsula, a rugged and roadless area of small lakes, ponds, bogs, and meadows.

The area has a rich, varied history. First inhabited by Native Americans who lived off the land, the area was crisscrossed in the late 18th and early 19th centuries by voyageurs in birch-bark canoes who worked primarily for fur trade companies. They were followed by miners, after gold was discovered on Little American Island in Rainy Lake in 1893. By 1910, the miners had deserted the unsuccessful mines, and loggers moved in to cut millions of white and red pine, spruce, and fir. By 1920, virtually all of the virgin timber had been cut and the loggers, too, moved on. Commercial fishing for sturgeon, walleye, northern pike, and whitefish, which began around the time of the gold rush, reached its peak in the 1930s. Now, this watery retreat has been left to anglers, campers, and boaters attracted by its splendid isolation.

The park is in the heart of the only region in the continental United States where the eastern timber wolf survives. Among national parks, it is the best bet in the lower 48 for seeing bald eagles, who nest along the shores. Since water covers one-third of the park's surface, aquatic animals such as beaver and otter thrive, as do water birds such as cormorants, kingfishers, great blue herons, and loons.

Boating is one of the major activities in the park. Boat tours are available by reservation, and houseboats can be rented. An angler's dream, the waters are world-renowned for walleye, northern pike, and smallmouth bass; ice fishing is also popular. Cross-country skiing, snowshoeing, and winter camping are popular from late December to late March. Snowmobile trails on the frozen lake surfaces are part of a regional network of trails. Ice roads also offer a unique way to experience Voyageurs. Check the park's website for current winter conditions, which can change rapidly, especially early and late in the season. Note that the first measurable snow here often occurs in October, and the last can be as late as May.

MN 60

MN 137

Voyageurs National Park

WATERTON–GLACIER

Going-to-the-Sun Mountain

Southwestern Alberta, Canada;
northwestern Montana
Established June 18, 1932
Glacier, 1,013,126 acres
Waterton Lakes, 124,800 acres

Headquarters
Glacier National Park
PO Box 128
West Glacier, MT 59936
406.888.7800
www.nps.gov/glac
@GlacierNPS

Waterton Lakes National Park
Box 200, Waterton Park, Alberta
Canada T0K 2M0   403.859.5133
www.pc.gc.ca/en/pn-np/ab/
waterton
@WatertonLakesNP

Visitor Centers
Waterton Information Centre
(403.859.5133); a new visitor centre
is under construction. In Glacier,
Logan Pass Visitor Center, on
Going-to-the-Sun Rd., and St. Mary
Visitor Center (406.732.7750), at St.
Mary Entrance, are open seasonally.
Apgar Visitor Center
(406.888.7800) at West Entrance is
open year-round, but only on week-
ends in the fall, winter, and spring.

Entrance Fees
Waterton: Adult, C$10.00, senior
citizen C$8.40, free admission
for youth 17 & under; family rate
C$20.00. Glacier, $35 summer/$25
winter per vehicle for 7 days;
$30/$20 per motorcycle; $20/$15
per pedestrian.

Accommodations
In Waterton, two campgrounds.
Townsite campground accepts
reservations (call 877.737.3783),
Belly River is temporarily closed,
normally open as first-come,
first-served May to September. In
Glacier, 13 campgrounds, 14-day
limit July 1 to Labor Day, most
open May/June to September. Reservations at Fish Creek and Many
Glacier (recreation.gov), others
first-come, first-served; three
campsites temporarily closed. A
range of other lodging in Glacier
is available; see the park's website
for details. For lodging in Waterton, call Waterton Lakes Chamber
of Commerce (mywaterton.ca,
403.859.2042).

When to Go
While winter draws intrepid cross-
country skiers from December to
June, Waterton–Glacier is best in
summer. The higher portions of
Going-to-the-Sun Rd. close by the
end of October to late May or June.
Higher hiking trails can remain
snowed in to mid-July, and snow
can fall unexpectedly in summer.

Though Glacier National Park
still holds 26 glaciers, it owes
its name to vast flows of ice
that carved, millennia ago,
what naturalist John Muir
called "the best care-killing
scenery on the continent."
He was likely not the first
to think so. Drawn by the
sheltered valleys and bountiful resources, people have
used these mountains for
more than 8,000 years. Early
native tribes tracked buffalo
across the plains and fished
the lakes; and the Blackfeet
controlled the land well into
the 19th century.

Rich in plant life and
wildlife, Glacier is one of
few parks where you can
still find grizzly bears and
gray wolves. Other fauna
include eagles, elk, mountain
lions, mink, and mountain
goats. Mackinaw trout and
northern pike may be fished
at Two Medicine Lake in

Two Medicine Valley, in the
park's southeastern section.
Affirming the fact that nature
transcends political boundaries, the eagles and elk
move freely between Glacier
and its Canadian neighbor,
Waterton Lakes National
Park – a fact which helped
inspire the creation of the
Waterton–Glacier Interna-
tional Peace Park in 1932.

Known for the sheer
walls and horned summits
of its tightly packed peaks,
Glacier is also home to more
than 700 stunning lakes and
thousands of miles of rivers
and streams. At the bottom
of stark cliffs sit forested
valley floors and tranquil
meadows. Of its six large
lakes, Lake McDonald is the
biggest, stretching ten miles
from the park's edge into its
core. During summer, runoff
from glaciers like Grinnell
cascades off the mountains,
plunging into the deep, ice-

cold lakes. With such awesome spectacles, it's little
wonder the Blackfeet called
the area sacred.

In addition to Glacier's
bounty of hiking trails, visitors
can enjoy the popular all-day
horseback trips that depart
from Many Glacier Valley.
For river lovers, raft floats of
anywhere from a half-day
to six-days' time explore the
north and middle forks of
the Flathead River. Glacier
also offers what many consider the best scenic drive in
America. The aptly named
Going-to-the-Sun Road
traverses the Continental
Divide on 50 miles of unfor-
gettable roadway, complete
with snowmelt spilling over
sheer cliffs, hairpin turns, and
17 scenic overlooks. Skirting
St. Mary Lake, on the park's
east side, the road twists up
to 6,646-foot Logan Pass,
where a popular trailhead
leads along a part-boardwalk
trail. Views include Triple
Divide Peak, where mountain
waters divide and head to
either the Arctic, the Atlantic,
or the Pacific Ocean. The road
descends to Lake McDonald,
where visitors can swim (on
warm days) and take boat
tours from a historic mountain lodge.

Other park highlights
include Red Eagle Lake, with
its beautiful falls and gorge.
A day-long scenic excursion
on Chief Mountain International Highway takes you
into Waterton Lakes National
Park. A popular hike, if you
have time, is Crypt Lake. The
10.8-mile trail leads past
four waterfalls and through
a 60-foot rock tunnel.

YELLOWSTONE

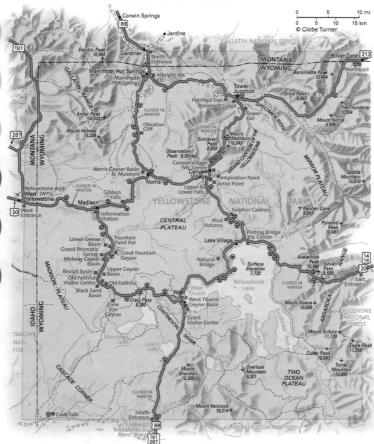

Canyon Visitor Education Center
(307.344.2550), open late spring to
mid-autumn. West Thumb Information Station (307.242.7690),
open late May to early October.
Fishing Bridge and Grant visitor
centers, temporarily closed.

Entrance Fees
$35 per vehicle for 7 days; $30 per
snowmobile or motorcycle; $20
per pedestrian, cyclist, or skier.

Accommodations
Seven campgrounds operated by
the National Park Service on a first-
come, first-served basis. Check park
website for opening dates; some
campsites are temporarily closed
for 2021. Five campgrounds operated by Yellowstone National Park
Lodges accept reservations; check
for latest opening schedule (yellow
stonenationalparklodges.com,
307.344.7311). Lodging ranges from
rustic cabins to a hotel with luxury
suites; most open seasonally, with
two also open in winter. For additional lodging options, contact
the West Yellowstone Chamber
of Commerce (destinationyellow-
stone.com, 406.646.7701).

When to Go
Most visitors see Yellowstone in
the summer, but a winter visit can
be just as breathtaking. In winter,
except for the route from the North
Entrance at Gardiner, MT, to the
Northeast Entrance at Cooke City,
MT, all roads are closed, making visits limited for all but snowmobilers.
Two lodges are open for winter visitors mid-December to early March.

Yellowstone is a land of
superlatives. The nation's
first national park has the
world's greatest concentration of geysers, mud pots,
fumaroles, and hot springs;
the largest number of free-
roaming wildlife in the lower
48, including bison, grizzly
bears, and reintroduced
gray wolves; more land than
Rhode Island and Delaware
combined; and one of North
America's largest mountain
bodies of water, Yellowstone
Lake. In 1872, President
Ulysses S. Grant convinced
Congress to set aside 3,472
square miles of land for
a national park, largely
because of the area's strange
and beautiful hydrothermal
wonders – in those days,
wildlife had no value, and
the land would not support
ranches or farms.

An early visitor, fur trap-
per Joe Meek, described the
whole area as "country that
was smoking with vapor
from boiling springs and

burning with gasses issu-
ing from small craters, each
of which was emitting a
sharp, whistling sound." Fur
trappers weren't the only
early visitors to Yellowstone.
Archaeological evidence
suggests the park has been
home to Native Americans
at least since the end of the
last ice age, 8,500 years ago.
Blackfeet, Crow, Shoshone,
and other tribes visited the
area, particularly in summer,
though the only permanent
residents were the Tukudikas.
Today, Yellowstone is host
to four million visitors each
year (more than saw the park
in the first 60 years of its
existence).

The park's major scenic
attractions are located along
the 142-mile Grand Loop
Road, the roughly figure-
eight-shaped road in the
center of the park. Many of
the most famous geysers
and hot springs are located
on the west side of the loop,
including Old Faithful, whose

eruption intervals range
from 58 to 104 minutes (the
current average interval is
94 minutes) and Fountain
Paint Pot, whose hot springs
vary in color depending on
the presence of bacteria and
algae, as well as the compo-
sition of the surrounding rock.
In the Upper Geyser Basin
north of Old Faithful, Giant
Geyser has rejuvenated after
40 years of near dormancy;
and nearby Grand Geyser
erupts every 5 to 8 hours in
spectacular bursts reaching
200 feet high.

On the east side of Grand
Loop Road, from Canyon Village north to Tower Junction,
is the Grand Canyon of the
Yellowstone, whose golden-
hued cliffs were created by
thermal water acting on
volcanic rock. Here, the Yellowstone River plunges 1,200
feet. A three-mile hike up
Mount Washburn, whose
slopes are carpeted with
wildflowers in June and
July, takes you to a summit
where, on a clear day, you can
view Yellowstone Lake and
the Tetons to the south and
the Beartooth Range to the
east. Yellowstone's Roosevelt
Country, at the northeast
top of the loop, is known
for its rolling hills covered
with sagebrush, fir, pine,
and aspen, and its sparkling
streams teeming with trout.
The southern end of the
loop includes Lake Country.
Formed by the forces of
volcanoes and glaciers,
Yellowstone Lake is a prime
habitat for a variety of birds
and mammals, as well as
being spectacular scenery.

**Lower Falls
of the Yellowstone River**

Northwestern Wyoming, with
portions extending into
southwestern Montana and
eastern Idaho
Established March 1, 1872
2,219,791 acres

Headquarters
Yellowstone National Park
PO Box 168
Yellowstone Nat'l Park, WY 82190

307.344.7381
www.nps.gov/yell
@YellowstoneNPS

Visitor Centers
Albright Visitor Center (307.344.2263),
in Mammoth Hot Springs, June–
December. Old Faithful Visitor Center
(307.344.2751), late May–December.

YOSEMITE

East-central California
Established October 1, 1890
761,748 acres

Headquarters
Yosemite National Park
PO Box 577
Yosemite, CA 95389
209.372.0200
www.nps.gov/yose
@YosemiteNPS

Visitor Centers
Yosemite Valley Visitor Center,
Yosemite Valley, open all year.
Wawona Visitor Center, open May
to October. Big Oak Flat Information Station, near park entrance at Hwy. 120, open May to October. Tuolumne Meadows Wilderness Center, just off Tioga Rd., open late May to late September, depending on seasonal closure of Tioga Pass. Note: check park website for temporary visitor center closures.

Entrance Fees
$35 per vehicle for 7 days; $30 per motorcycle, $20 per pedestrian or bus passenger 16 and older.

Accommodations
For 2021, reservations are required at all campsites (recreation.gov, 877.444.6777). In Yosemite Valley, Upper Pines open all year; Lower Pines, North Pines, and Camp 4, late April/May to late October. South of the valley, Wawona and Bridalveil Creek are open mid-June to Sept/Oct. North of the valley, six campsites are temporally closed for 2021; Tuolumne Meadows is open mid-July to late September. The Ahwahnee Hotel and Yosemite Valley Lodge open all year; other lodging options in the park open seasonally. For reservations in Yosemite, contact Yosemite Hospitality (www.travelyosemite.com, 888.413.8869).

When to Go
Summer draws the heavy traffic to Yosemite, but late spring, when its famous waterfalls are usually at their peak, or autumn, when oaks turn golden and herds of wild deer migrate through the valley, are the best times for a visit. Avoid holiday weekends and expect filled campgrounds June to August. In winter, Yosemite Valley and Wawona are accessible by auto, but the Tioga Road is closed, usually beginning in November. The Glacier Point/Badger Pass Road is plowed to the Badger Pass Ski Area for access to downhill and cross-country skiing.

El Capitan and Merced River

"No temple made with hands can compare with Yosemite," wrote John Muir, the man hailed as the founder of the national park system. To this temple come more than four million adoring fans every year. They come for its glacially polished granite cliffs, alpine meadows, and groves of giant sequoias. But most of all they come to stand and gape at its stunning waterfalls, like Yosemite Falls, which, at 2,425 feet, is one of the highest waterfalls in North America (taller than the Sears and Eiffel Towers combined).

Muir was so enamored with Yosemite that this peripatetic traveler said he was "willing to stay forever in [Yosemite] like a tree." The Ahwahneechee, who named the land Ahwahnee, meaning wide-gaping mouth or deep, grassy valley, were one group of early residents, fishing the streams, gathering black oak acorns, and hunting the plentiful game in summer. It remained largely unvisited by interlopers until 1851, when the U. S. Cavalry's Mariposa Battalion tracked renegade Indians into the valley.

In 1864, President Lincoln made Yosemite a protected reserve; in 1890 it became a national park. Ever since, it has drawn throngs of visitors, anxious to see for themselves one of the great natural wonders of the world. But beauty and popularity come at a cost: Some say Yosemite is being loved to death, particularly in summer, when the valley floor, which comprises only seven square miles of the park's 1,190-square-mile area, draws more than 90 percent of the visitors. On busy weekends, one-way traffic between major attractions resembles the rush-hour scene in Los Angeles. Plans are slowly being implemented to "de-develop" Yosemite.

There are two simple tricks to getting the most out of a trip to Yosemite. First, plan a visit in early spring or late fall – only around a quarter of Yosemite's guests arrive between November and April. Second, take the time to explore attractions beyond the heavily visited valley. The magnificent Mariposa Grove of Giant Sequoias lies near the park's south entrance, 37 miles from the valley. Here, you'll find 200-foot-tall giant sequoias that are among the largest and oldest of all living things. At Glacier Point you'll get the most spectacular view of the entire valley, from Yosemite Falls on the north wall of the valley to the sheared granite wall of Half Dome, looming nearly 5,000 feet above the valley's east end. In winter, the road to the point is closed at Badger Pass Ski Area; but Glacier Point is also a favorite destination of cross-country skiers. Some of the most rugged, sublime alpine scenery can be found in the Tuolumne Meadows and high country of the eastern part of the park. The Tioga Road, open in summer, offers a 39-mile scenic drive through an area of sparkling lakes, delicate meadows, and lofty peaks that was once under glacial ice.

CA 13
CA 134

ZION

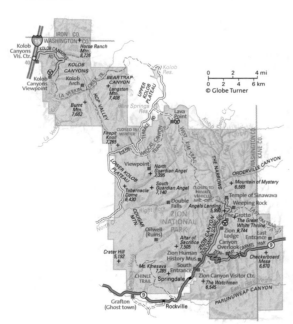

Southwestern Utah
Established November 19, 1919
147,243 acres

Headquarters
Zion National Park
1 Zion Park Blvd., State Rte. 9,
Springdale, UT 84767
435.772.3256
www.nps.gov/zion
@ZionNPS

Visitor Centers
Zion Canyon Visitor Center
(435.772.3256), near southern entrance on Utah Rte. 9, and Kolob Canyons Visitor Center, northwest corner of park via Exit 40 off I-15, open all year.

Entrance Fees
$35 per vehicle; $30 per motorcycle; $20 per person on foot or bicycle. All fees valid for 7 days.

Accommodations
Three campgrounds, all with 14-day limits March to November. Reservations needed (877.444.6777) at Watchman, open all year, and South, open mid-March to November; Lava Point (first-come, first-served) open from May to September, depending on weather. Lodging available at Zion Lodge (zionlodge.com, 888.297.2757), off Utah Rte. 9. For nearby lodging outside the park, contact Zion Canyon Visitors Bureau (www.zionpark.com).

When to Go
The park is busiest April to October. Mild spring and fall temperatures make hiking more pleasant, and rain showers moderate the summer heat, rolling in spectacularly and creating new waterfalls on the sheer cliffs. Snow closes higher hiking trails in winter. Zion Canyon Scenic Drive is closed to private vehicles from March through November; during this time a free shuttle service is available to transport visitors between various points in Zion Canyon. Winter conditions in the park are cold and often wet. Roads are plowed, but some trails may close due to winter hazards. Check the park website for current weather conditions.

Mormon pioneers who first saw the sculpted rocks and multicolored walls rising above the Virgin River in Utah's high-plateau country named the area Zion, after the heavenly city of God. Thereafter, a Methodist pastor, the Reverend Frederick Fisher, on an expedition to the canyon in 1916 also gave religious names to the more spectacular natural wonders, like the towering 2,400-foot monolith called Great White Throne or the Three Patriarchs (Abraham, Isaac, and Jacob), sheer faces carved by wind and water from Navajo sandstone. Zion is, indeed, a bit of heaven on earth, a park with such magnificent scenery that it has been variously referred to as Yosemite in color and as a vertical little Grand Canyon.

The highlight of the park is Zion Canyon, a half-mile-deep slash formed by the Virgin River cutting through the sandstone. Some say the main drive to view the area is among the most spectacular in the country, a bit like driving through the bottom of the Grand Canyon. The Zion Canyon Scenic Drive is a narrow paved road with colorful vistas of looming cliffs, domes, and mountains, ending at the Temple of Sinawava, named for the Paiute wolf god or good spirit. Another fork road into the park, a stretch of Utah Highway 9 called the Zion–Mount Carmel Highway, passes through mile-long Zion Tunnel, 800 feet above the canyon floor, where windowlike galleries gouged in the rock let you look out on scenic wonders like East Temple and the 400-foot-high Great Arch. In summer, only shuttles are allowed on the Zion Canyon Scenic Drive. Tickets are not required, and the shuttles offer an easy way to view the stunning scenery in Zion Canyon.

The best way to experience the park, though, is by going on a hike. Zion has an extensive trail system and a wide range of choices. Weeping Rock Trail is one of three easy, self-guided nature trails. The less than half-mile-long hike climbs a hundred feet to a lovely area of hanging gardens. Another popular trail is the Riverside Walk, an easy two-mile round-trip ramble that begins at the end of the Zion Canyon Scenic Drive and takes you past golden columbine and stands of shady cottonwood and ash. More difficult is the Angels Landing Trail, a five-mile round-trip workout with steep drop-offs, a 1,500-foot elevation gain, and one of the best overall views of the park.

There are 262 species of birds in the park, making it a favorite location for birdwatching. Zion is known for southern birds at the northern limit of their range, including gray vireos and the common black hawk.

The peaceful Virgin River

UT 109
UT 141

UNITED STATES

© Globe Turner

	ANNISTON	AUBURN	BIRMINGHAM	CHATTANOOGA, TN	COLUMBUS, GA	DECATUR	DEMOPOLIS	DOTHAN	FLORENCE	GADSDEN	HAMILTON	HUNTSVILLE	MERIDIAN, MS	MOBILE	MONTGOMERY	OPP	SELMA	TUSCALOOSA
BIRMINGHAM	66	141		149	167	83	120	191	121	63	92	101	149	258	88	166	94	61
DOTHAN	207	125	191	311	97	273	198		310	251	282	291	246	199	103	66	147	237
HUNTSVILLE	100	241	101	109	266	25	215	291	65	74	104		245	357	187	265	194	156
MOBILE	279	227	258	403	252	340	144	199	377	317	286	357	132		173	146	194	205
MONTGOMERY	106	54	88	234	79	170	100	103	207	148	191	187	153	173		81	51	134

DRIVING DISTANCES IN MILES   SEE ALSO MILEAGE AND DRIVING TIME MAP ON PAGE 144

DRIVING DISTANCES IN MILES	FAIRBANKS	GLENNALLEN	HAINES	HOMER	JUNEAU	KENAI	KETCHIKAN	SEWARD	SKAGWAY	TOK	VALDEZ
ANCHORAGE	378	184	760	225	841*	161	1608*	129	807	323	301
FAIRBANKS	378		249	645	603	726*	539	1493*	507	691	207	366
HOMER	225	603	408	985		1066*	87	1833*	172	1031	547	525
SKAGWAY	807	691	623	27	1031	101*	968	954*	935		484	740
TOK	323	207	139	438	547	518*	484	1258*	451	484		256

*DISTANCE INCLUDES FERRY TRAVEL

DENALI NATIONAL PARK AND PRESERVE ALASKA

FAIRBANKS

JUNEAU

ANCHORAGE

ALEUTIAN ISLANDS

Distances in the U.S. shown in miles.
Distances in Canada shown in kilometers.

© Globe Turner

	BULLHEAD CITY	CASA GRANDE	CHINLE	DOUGLAS	FLAGSTAFF	GRAND CANYON	HOLBROOK	KINGMAN	LAKE HAVASU CITY	NOGALES	PAGE	PHOENIX	PRESCOTT	SAFFORD	SHOW LOW	TUCSON	WICKENBURG	YUMA
FLAGSTAFF	180	188	216	374		89	93	148	209	318	135	137	89	271	140	255	147	320
KINGMAN	34	235	364	421	148	175	240		60	365	281	184	150	353	288	302	134	215
PHOENIX	217	50	353	237	137	226	230	184	193	181	272		96	169	178	118	51	183
TUCSON	335	68	366	120	255	345	240	302	311	65	390	118	214	128	193		169	241
YUMA	222	179	536	360	320	409	413	215	155	304	455	183	213	368	352	241	170	

DRIVING DISTANCES IN MILES   SEE ALSO MILEAGE AND DRIVING TIME MAP ON PAGE 144

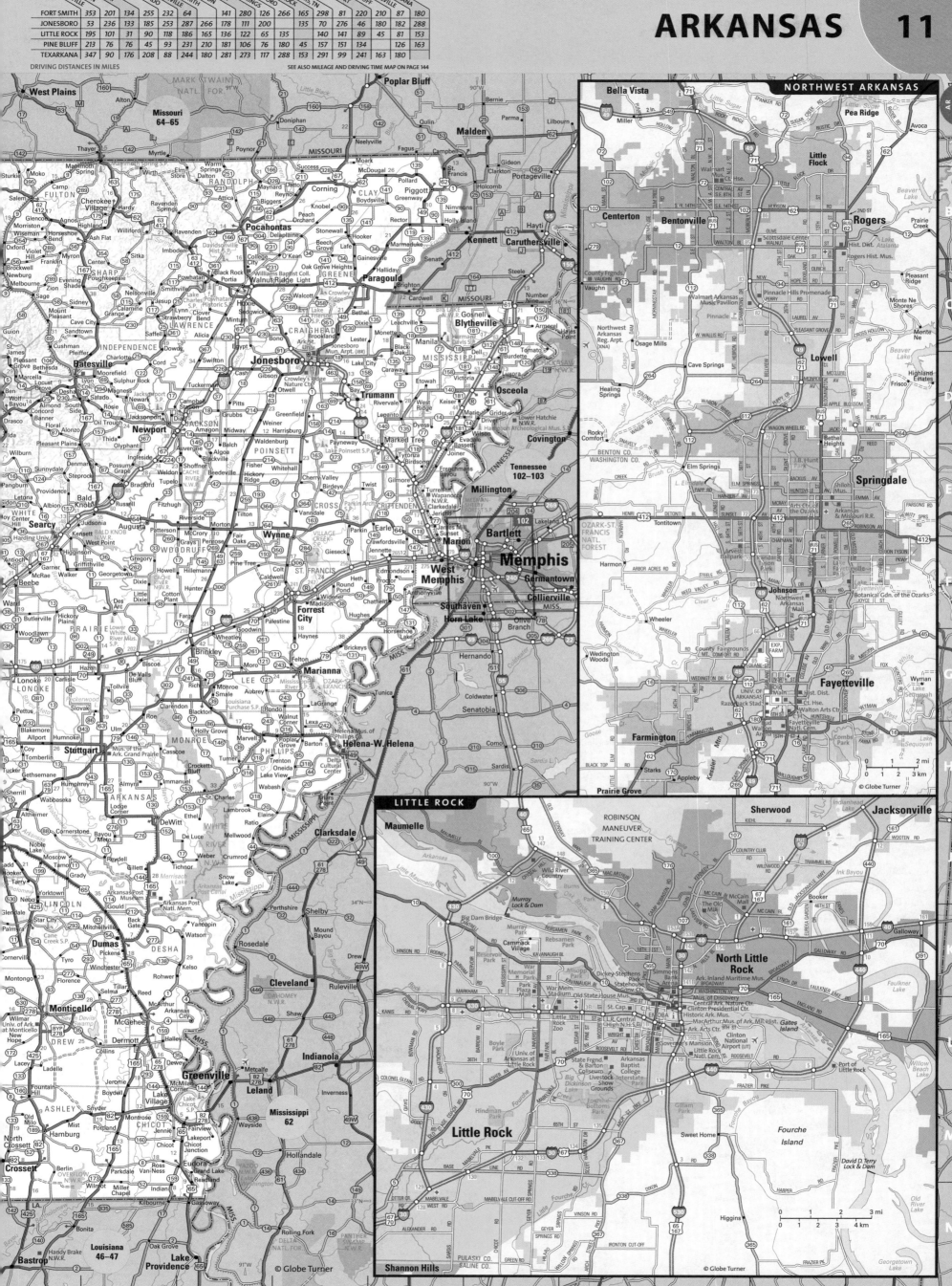

	CAMDEN	BLYTHEVILLE	CONWAY	DUMAS	EL DORADO	FAYETTEVILLE	FORT SMITH	HARRISON	HELENA	HOT SPRINGS	JONESBORO	LITTLE ROCK	MEMPHIS, TN	MENA	NEWPORT	PINE BLUFF	RUSSELLVILLE	TEXARKANA
FORT SMITH	353	201	134	255	232	64		141	280	126	266	165	298	81	220	210	87	180
JONESBORO	53	236	133	185	253	287	266		178	111	200		135	70	276	46	180	182	288
LITTLE ROCK	195	101	31	90	118	186	165	136	122	65	135		140	141	89	45	81	153
PINE BLUFF	213	76	76	45	93	231	210	181	106	76	180	45	157	151	134		126	163
TEXARKANA	347	90	176	208	88	244	180	281	273	117	288	153	291	99	241	163	180	

DRIVING DISTANCES IN MILES   SEE ALSO MILEAGE AND DRIVING TIME MAP ON PAGE 144

NORTHWEST ARKANSAS

LITTLE ROCK

© Globe Turner

MI 25 50

KM 25 50

MONTEREY BAY

PACIFIC OCEAN

	BISHOP	CHICO	EUREKA	FRESNO	MERCED	MONTEREY	NAPA	OAKLAND	REDDING	SACRAMENTO	SAN FRANCISCO	SAN JOSE	SANTA ROSA	SOUTH LAKE TAHOE	STOCKTON	SUSANVILLE	UKIAH	YOSEMITE VILLAGE
EUREKA	537	186		444	385	380	235	262	133	278	263	306	208	379	336	247	148	436
REDDING	426	74	133	344	284	323	193	213		166	222	250	228	266	214	114	193	336
SACRAMENTO	260	88	278	178	118	188	58	78	166		87	115	93	100	48	183	153	170
SAN FRANCISCO	283	182	263	190	131	114	47	9	222	87		43	56	185	82	277	116	183
SOUTH LAKE TAHOE	179	165	379	267	208	286	156	176	266	100	185	213	191		46	142	251	180

DRIVING DISTANCES IN MILES

SEE ALSO MILEAGE AND DRIVING TIME MAP ON PAGE 144

SACRAMENTO

STOCKTON

FRESNO

LAKE TAHOE

YOSEMITE NATIONAL PARK

KINGS CANYON–SEQUOIA NATL. PARKS

SANTA ROSA–NAPA VALLEY

	BARSTOW	BAKERSFIELD	BISHOP	EL CENTRO	FRESNO	LAS VEGAS	LOS ANGELES	MERCED	MONTEREY	PALM SPRINGS	SAN BERNARDINO	SAN DIEGO	SAN FRANCISCO	SAN JOSE	SAN LUIS OBISPO	SANTA BARBARA	VISALIA	YOSEMITE VILLAGE
Fresno	111	246	219	451		399	219	59	158	324	277	342	190	153	134	258	43	90
Los Angeles	111	118	270	234	219	274		278	327	110	62	124	385	342	192	97	183	308
Monterey	229	365	315	559	158	517	327	122		438	385	450	114	74	149	242	204	201
San Diego	234	181	366	117	342	337	124	401	450	143	111		508	465	314	221	306	431
San Francisco	287	423	283	617	190	575	385	131	114	490	443	508		43	234	327	233	183

DRIVING DISTANCES IN MILES

SEE ALSO MILEAGE AND DRIVING TIME MAP ON PAGE 144

© Globe Turner

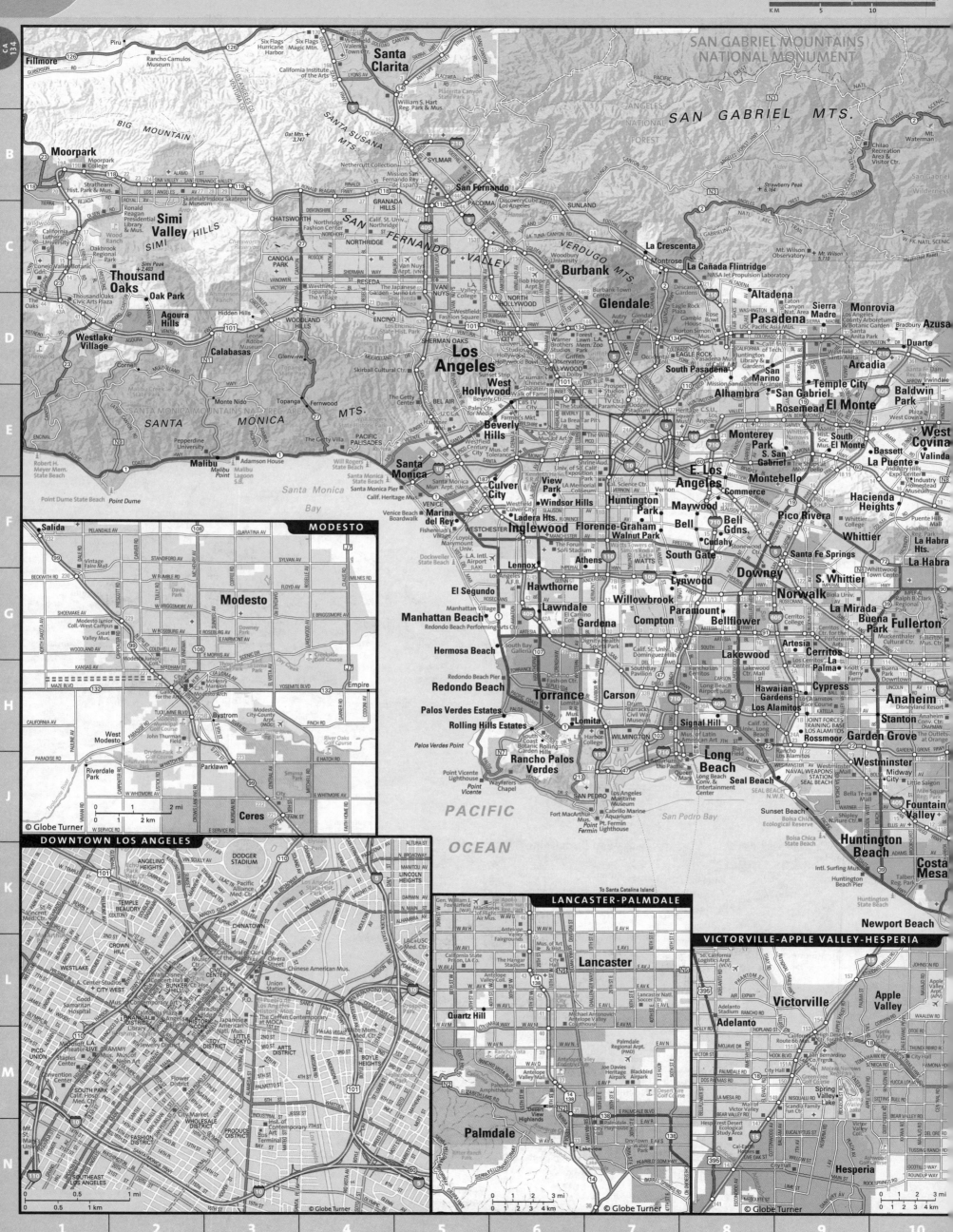

MI 5 10
KM 5 10

SAN GABRIEL MOUNTAINS NATIONAL MONUMENT

Fillmore
Piru
Six Flags Hurricane Harbor
Westfield Valencia Town Center
Rancho Camulos Museum
California Institute of the Arts
Santa Clarita
SAN GABRIEL MTS.
Mt. Waterman
Chilao Recreation Area & Visitor Ctr.

Moorpark
Moorpark College
Stratheam Hist. Park & Mus.
BIG MOUNTAIN
SANTA SUSANA MTS.
William S. Hart Reg. Park & Mus.
Strawberry Peak
San Gabriel Wilderness

Simi Valley
SIMI HILLS
Ronald Reagan Presidential Library & Mus.
Oat Mtn. 3,747
San Fernando
SYLMAR
PACOIMA
Discovery Cube Los Angeles
La Crescenta
Montrose
Mt. Wilson Observatory
Mt. Wilson 5,710

Thousand Oaks
California Lutheran University
Wood Ranch
Oakbrook Regional Park
Simi Peak 2,403
CHATSWORTH
GRANADA HILLS
Burbank
La Cañada Flintridge
NASA Jet Propulsion Laboratory
Altadena
Sierra Madre
Monrovia
Arcadia
Azusa

Oak Park
Hidden Hills
CANOGA PARK
NORTHRIDGE
Calif. St. Univ. Northridge
NORTH HOLLYWOOD
Glendale
Pasadena
Rose Bowl
South Pasadena
Duarte

Agoura Hills
Calabasas
WOODLAND HILLS
RESEDA
SHERMAN OAKS
STUDIO CITY
Griffith Observatory
Eagle Rock
Alhambra
San Gabriel
San Marino
Temple City
Baldwin Park

Westlake Village
Glenview
ENCINO
Los Angeles
West Hollywood
Hollywood
Dolby Theatre
Rosemead
El Monte
West Covina

SANTA MONICA MOUNTAINS NAT. REC. AREA
The Getty Center
BEL AIR
U.C.L.A.
Beverly Hills
Monterey Park
S. San Gabriel
Bassett
La Puente
Valinda

Malibu
Adamson House
Malibu Lagoon S.B.
PACIFIC PALISADES
The Getty Villa
Santa Monica
Culver City
View Park
E. Los Angeles
Montebello
Commerce
Hacienda Heights

Point Dume State Beach
Point Dume
SANTA MONICA MTS.
Topanga
Fernwood
VENICE
Marina del Rey
Ladera Hts.
Windsor Hills
Huntington Park
Maywood
Bell
Bell Gdns.
Cudahy
Pico Rivera
Whittier
Whittier College
La Habra Hts.

Santa Monica Bay

Salida
Ceres
Modesto
Empire
Bystrom
West Modesto
Riverdale Park
Parklawn

Inglewood
Florence-Graham
Walnut Park
South Gate
Downey
S. Whittier
La Habra

Dockweiler State Beach
L.A. Intl. Airport (LAX)
Lennox
Athens
Watts
Lynwood
Norwalk
La Mirada
Buena Park
Fullerton

El Segundo
Hawthorne
Willowbrook
Paramount
Bellflower
Artesia
Cerritos
La Palma

Manhattan Beach
Lawndale
Gardena
Compton
Lakewood
Cypress
Anaheim
Stanton

Hermosa Beach
Redondo Beach
Torrance
Carson
Hawaiian Gardens
Los Alamitos
Rossmoor
Garden Grove
Westminster

Palos Verdes Estates
Signal Hill
Long Beach
Seal Beach
Fountain Valley

Rolling Hills Estates
Palos Verdes Point
Rancho Palos Verdes
Point Vicente
San Pedro
Huntington Beach

PACIFIC OCEAN
San Pedro Bay
Costa Mesa

Newport Beach

DOWNTOWN LOS ANGELES

DODGER STADIUM
ELYSIAN HEIGHTS
LINCOLN HEIGHTS
CHINATOWN
Union Station
CIVIC CENTER
LITTLE TOKYO
BOYLE HEIGHTS
WESTLAKE
FINANCIAL DISTRICT
SOUTH PARK
FASHION DISTRICT
PICO UNION

LANCASTER-PALMDALE

Lancaster
Quartz Hill
Palmdale

VICTORVILLE-APPLE VALLEY-HESPERIA

Victorville
Apple Valley
Adelanto
Hesperia

© Globe Turner

	ASPEN	BOULDER	BURLINGTON	COLORADO SPRINGS	CRAIG	DENVER	DURANGO	ESTES PARK	FORT COLLINS	GLENWOOD SPRINGS	GRAND JUNCTION	GREELEY	LAMAR	MONTROSE	PUEBLO	STERLING	TRINIDAD
COLORADO SPRINGS	162	157	97		152		270	70	314	134	133	226	318	133	161	236	43	194	127
DENVER	230	164	27	168	70		203		337	64	64	158	250	64	208	277	111	130	196
DURANGO	152	244	366	461	314	321	337			402	399	226	169	399	354	107	271	465	260
GRAND JUNCTION	261	135	254	418	318	152	250	169	291		311	92		311	458	62	360	377	444
PUEBLO	215	185	139	191	43		312	111	271	175	175	268	360	175	203	229		236	84

DRIVING DISTANCES IN MILES

SEE ALSO MILEAGE AND DRIVING TIME MAP ON PAGE 144

© Globe Turner

	BRIDGEPORT	DANBURY	HARTFORD	MERIDEN	MIDDLETOWN	NEW HAVEN	NEW LONDON	NORWICH	PROVIDENCE, R.I.	PUTNAM	SPRINGFIELD, MA.	STAMFORD	STORRS	TORRINGTON	WATERBURY	WILLIMANTIC	WINDSOR LOCKS
BRIDGEPORT		31	56	37	44	19	64	60	72	118	100	81	21	75	54	33	79	68
HARTFORD	56	57		21	16	39	46	115	38	73	46	23	17	21	25	30	25	13
NEW LONDON	64	81	46		50	40	46		124	15	58	51	71	85	41	72	65	29	59
TORRINGTON	54	40	25	40	40	50	72	107	64	98	71	50	74		46	21	51	38
WATERBURY	33	31	30	20	25	30	65	99	60	118	76	55	53	51	21		56	43

DRIVING DISTANCES IN MILES   SEE ALSO MILEAGE AND DRIVING TIME MAP ON PAGE 144

CONNECTICUT

NEW LONDON

HARTFORD

© Globe Turner

DELAWARE

DRIVING DISTANCES IN MILES

SEE ALSO MILEAGE AND DRIVING TIME MAP ON PAGE 144

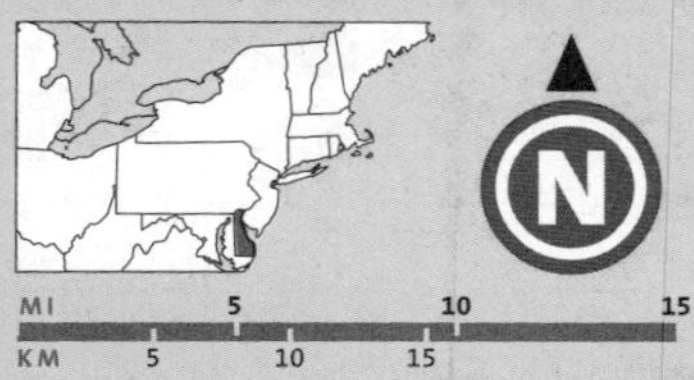

	BETHANY BEACH	DOVER	GEORGETOWN	HARRINGTON	LEWES	MILFORD	NEWARK	PHILADELPHIA, PA	REHOBOTH BEACH	SALISBURY, MD	SEAFORD	WILMINGTON
DOVER	55		38	17	42	21	41	74	43	56	36	44
NEWARK	94	41	78	57	81	60		44	83	97	76	14
REHOBOTH BEACH	13	43	18	31	10	24	83	116		46	35	86
SEAFORD	37	36	17	19	34	26	76	110	35	21		80
WILMINGTON	98	44	81	61	85	64	14	30	86	101	80	

WILMINGTON

NEWARK

DOVER

© Globe Turner

N

MI 5 10 15

KM 5 10 15

DOWNTOWN MIAMI

GULF OF MEXICO

ATLANTIC OCEAN

BISCAYNE NATIONAL PARK

© Globe Turner

	DAYTONA BEACH	FORT MYERS	FORT PIERCE	FORT WALTON BEACH	GAINESVILLE	JACKSONVILLE	LAKE CITY	LAKELAND	MELBOURNE	MIAMI	OCALA	ORLANDO	PANAMA CITY	PENSACOLA	ST. AUGUSTINE	TALLAHASSEE	TAMPA	TITUSVILLE
JACKSONVILLE	91	295	223	328	70		62	197	175	345	101	141	267	363	41	166	196	133
ORLANDO	56	155	120	425	117	141	157	56	72	232	80		345	460	103	262	82	40
PENSACOLA	455	589	567	39	349	363	306	471	526	681	383	460	102		403	200	474	497
TALLAHASSEE	258	392	369	166	152	166	109	274	329	483	186	262	104	200	207		277	300
TAMPA	138	123	172	440	132	196	172	37	142	274	95	82	474	185	277	121		

DRIVING DISTANCES IN MILES

SEE ALSO MILEAGE AND DRIVING TIME MAP ON PAGE 144

GAINESVILLE

DAYTONA BEACH

MELBOURNE–TITUSVILLE

© Globe Turner

	FORT LAUDERDALE	FORT MYERS	FORT PIERCE	KEY WEST	LAKELAND	MELBOURNE	MIAMI	ORLANDO	ST. PETERSBURG	SARASOTA	TAMPA	WEST PALM BEACH
FORT MYERS	139		126	308	113	178	155	155	108	74	123	125
FORT PIERCE	102	126		288	122	57	120	197	150	172	57	
MIAMI	23	155	122	168	236	179		232	259	225	274	67
ORLANDO	216	155	120	398	56	72	232		107	130	82	169
TAMPA	257	123	172	426	37	142	274	82	25	60		223

DRIVING DISTANCES IN MILES

SEE ALSO MILEAGE AND DRIVING TIME MAP ON PAGE 144

PORT ST. LUCIE

FORT MYERS

KEY WEST

Acworth
Kennesaw
Alpharetta
Johns Creek
Roswell
Duluth
Sandy Plains
Peachtree Corners
Due West
Sandy Springs
Dunwoody
Norcross
Marietta
Fair Oaks
Smyrna
Doraville
Powder Springs
Chamblee
Tucker
Lilburn
North Atlanta
Brookhaven
Mableton
Austell
Druid Hills
N. Druid Hills
Clarkston
Stone Mountain
Scottdale
Buckhead
N. Decatur
Avondale Estates
Grove Park
Decatur
Belvedere Park
Atlanta
East Atlanta
Candler-McAfee
Douglasville (part)
Gresham Park
Panthersville
Stonecrest
Ben Hill
East Point
Lakewood
Hapeville
South Fulton
College Park
Forest Park
Hartsfield-Jackson Atlanta International Airport (ATL)
Red Oak
Chattahoochee Hills
Union City
Riverdale
Fairburn
Palmetto

Albany

Country Club Estates
Dock Junction
Brunswick
St. Simons Island

DOWNTOWN ATLANTA

MIDTOWN
ENGLISH AVENUE
CENTENNIAL PLACE
CENTRAL PARK
VINE CITY
CASTLEBERRY HILL
DOWNTOWN
OLD FOURTH WARD
SWEET AUBURN
MECHANICSVILLE
SUMMERHILL
GRANT PARK

© Globe Turner

AUGUSTA

ATHENS

SOUTH CAROLINA

ALABAMA

TENNESSEE

NORTH CAROLINA

N.C.

TENN.

S.C.

Atlanta
Augusta
Athens
Macon
Columbus
Chattanooga
Rome
Dalton
Gainesville
Marietta
Smyrna
Roswell
Sandy Springs
Alpharetta
Duluth
Snellville
Lawrenceville
Winder
Monroe
Covington
McDonough
Griffin
Fayetteville
Peachtree City
Newnan
LaGrange
Carrollton
Douglasville
Cartersville
Calhoun
Cleveland
LaFayette
Cedartown
Fort Payne
Phenix City
Auburn
Opelika
Tuskegee
Warner Robins
Fort Valley
Perry
Dublin
Sandersville
Milledgeville
Eatonton
Greensboro
Washington
Thomson
Evans
Belvedere
North Augusta
Martinez
Aiken
Barnwell
Orangeburg
Walterboro
Beaufort
Burton
Statesboro
Swainsboro
Vidalia
Waynesboro
Sylvania
Millen
Greenville
Spartanburg
Greer
Taylors
Mauldin
Simpsonville
Easley
Seneca
Clemson
Anderson
Hartwell
Elberton
Abbeville
Greenwood
Laurens
Hendersonville
Brevard
Commerce
Toccoa

	ALBANY	AMERICUS	ATHENS	ATLANTA	AUGUSTA	BAINBRIDGE	BRUNSWICK	CHATTANOOGA TN	COLUMBUS	DUBLIN	GAINESVILLE	LA GRANGE	MACON	ROME	SAVANNAH	STATESBORO	VALDOSTA	WAYCROSS
ATLANTA	180	129	70		149	236	308	113	106	139	56	69	84	66	249	211	228	253
AUGUSTA	226	206	97	149		282	194	266	249	95	136	212	123	219	135	81	274	184
MACON	102	83	89	84	123	159	225	201	95	55	142	114		154	165	127	151	159
SAVANNAH	246	226	225	249	135	248	78	366	244	114	307	279	165	319		53	168	106
VALDOSTA	90	119	239	228	274	80	120	346	183	139	287	226	151	298	168	173		62

DRIVING DISTANCES IN MILES   SEE ALSO MILEAGE AND DRIVING TIME MAP ON PAGE 144

	HĀNA	HILO	HONOLULU	HO'OLEHUA	KAHULUI	KAILUA	KAILUA-KONA	LAHAINA	LĀNA'I CITY	LĪHU'E	WAILUA	WAIMEA
HILO	149*		217*	169*	121*	235*	88	142*	155*	319*	234*	54
HONOLULU	129*	217*		54*	101*	14	185*	92*	74*	102*	23	172*
KAHULUI	42	121*	101*	76		119*	109*	23	57	202*	118*	79*
KAILUA-KONA	137*	88	185*	157*	109*	203*		132*	143*	285*	202*	39
LĪHU'E	230*	319*	102*	156*	202*	120*	285*	225*	176*		119*	174*

DISTANCES IN MILES   * DISTANCE INCLUDES AIR TRAVEL

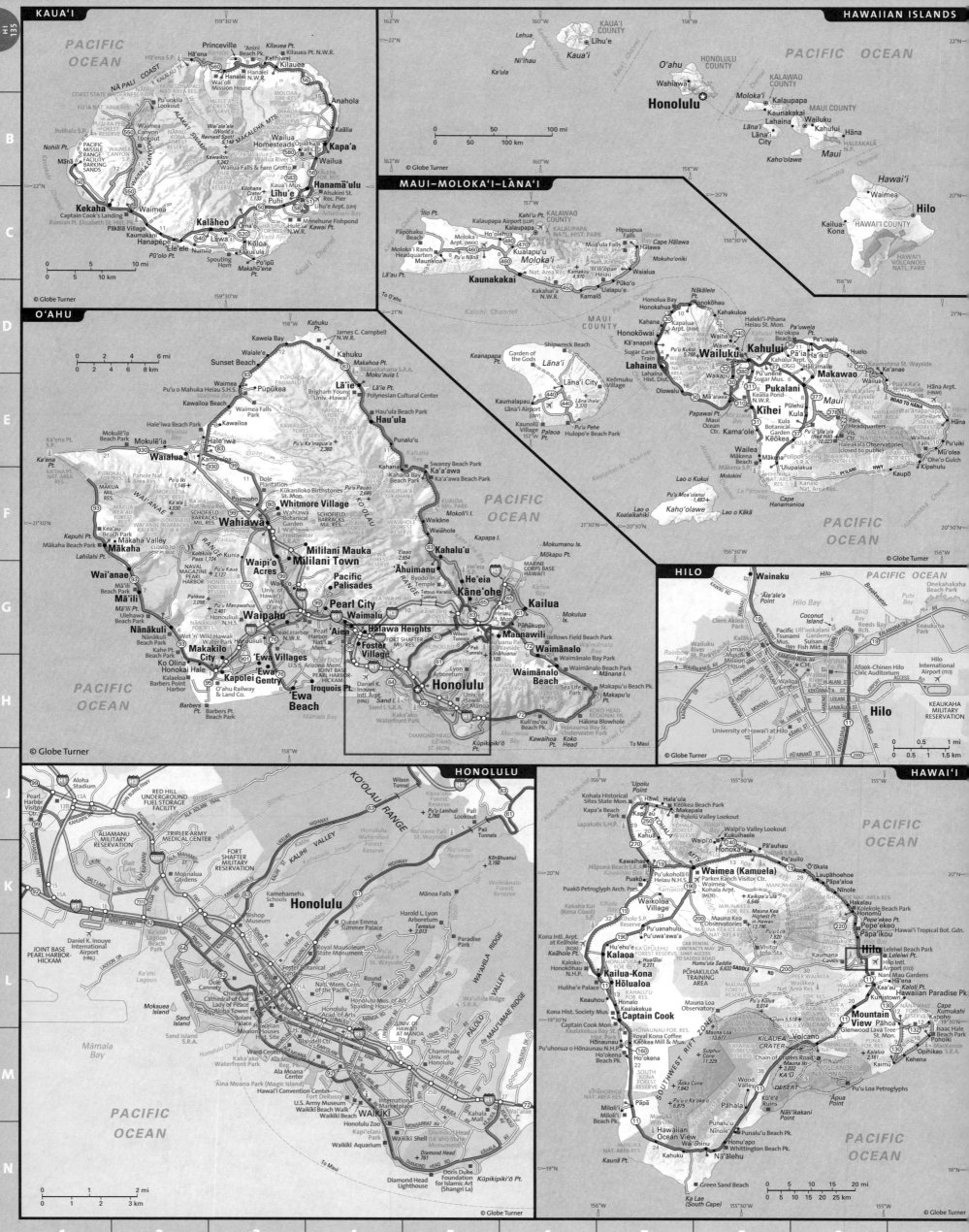

DRIVING DISTANCES IN MILES

SEE ALSO MILEAGE AND DRIVING TIME MAP ON PAGE 144

	BOISE	COEUR D'ALENE	GRANGEVILLE FALLS	IDAHO FALLS	KETCHUM	LEWISTON	MISSOULA, MT	MOUNTAIN HOME	POCATELLO	SALMON	SANDPOINT	TWIN FALLS
BOISE		406	202	288	163	270	374	49	241	247	452	134
COEUR D'ALENE	406		186	476	485	118	167	499	526	307	48	584
IDAHO FALLS	288	476	483		153	532	311	240	53	168	523	162
LEWISTON	270	118	74	532	477		221	363	555	337	166	448
POCATELLO	241	526	440	53	190	555	360	193		217	572	116

BOISE

POCATELLO

IDAHO FALLS

© Globe Turner

N

MI 10 20 30
KM 10 20 30

LAKE MICHIGAN

EASTERN TIME ZONE
CENTRAL TIME ZONE

INDIANA 38-39

WISCONSIN 118-41

IOWA 40-41

MISSOURI 64-65

ROCKFORD

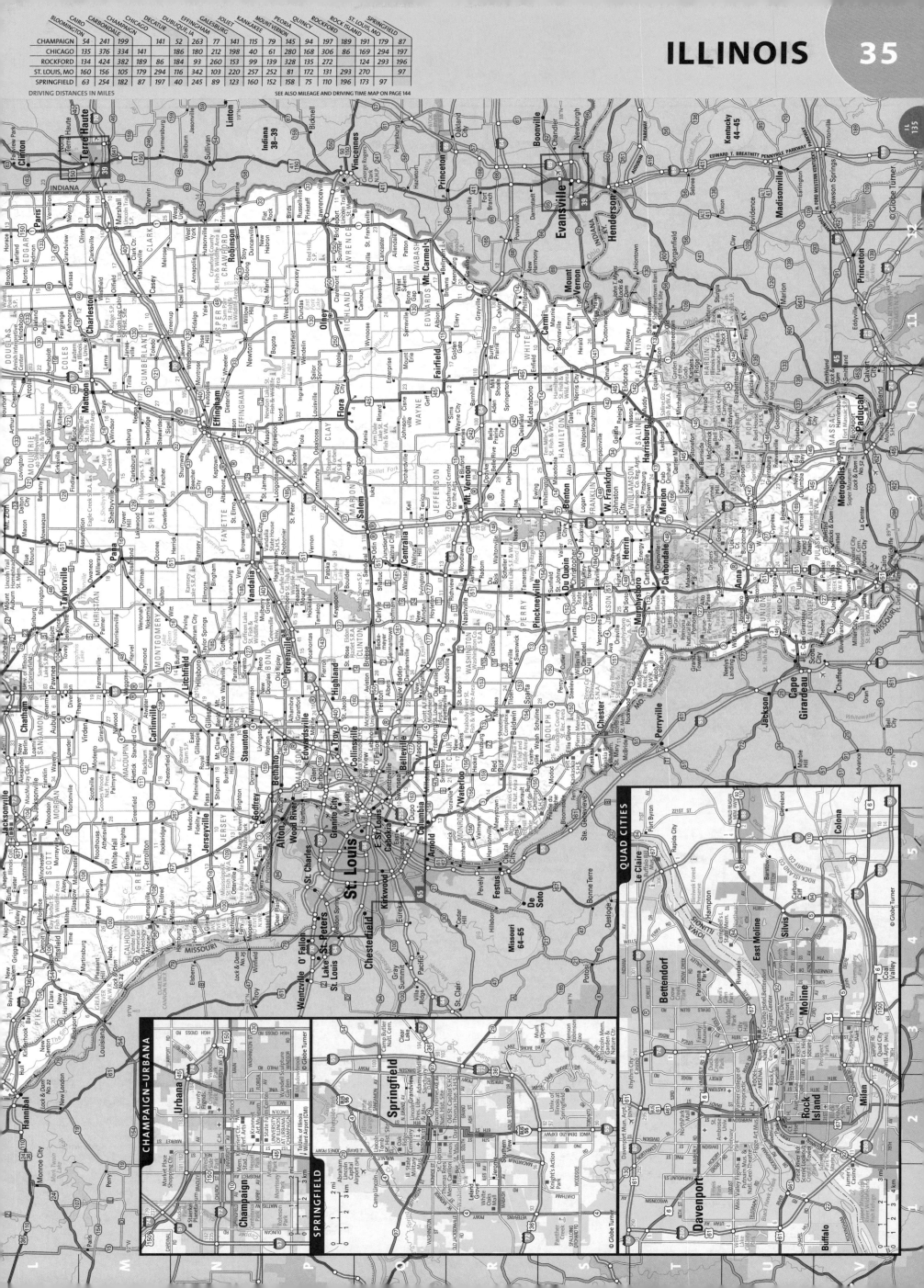

© Globe Turner

LAKE

MICHIGAN

IN
136

INDIANA

MI 10 20
KM 10 20
N

Indianapolis
Fort Wayne
South Bend
Chicago
Gary
Kokomo
Muncie
Anderson
Lafayette
West Lafayette
Elkhart
Mishawaka
Valparaiso
Logansport
Marion
Richmond
Danville
Joliet
Ohio 86–89
Michigan 56–57
Illinois 34–35
OHIO
MICHIGAN
ILLINOIS
LAKE MICHIGAN
EASTERN TIME ZONE
CENTRAL TIME ZONE
FORT WAYNE
ELKHART
LAFAYETTE
© Globe Turner

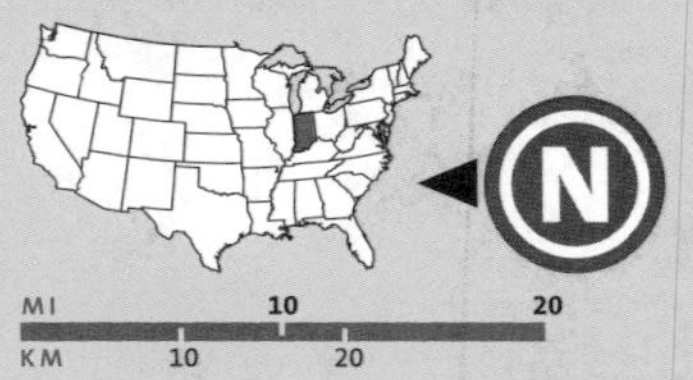

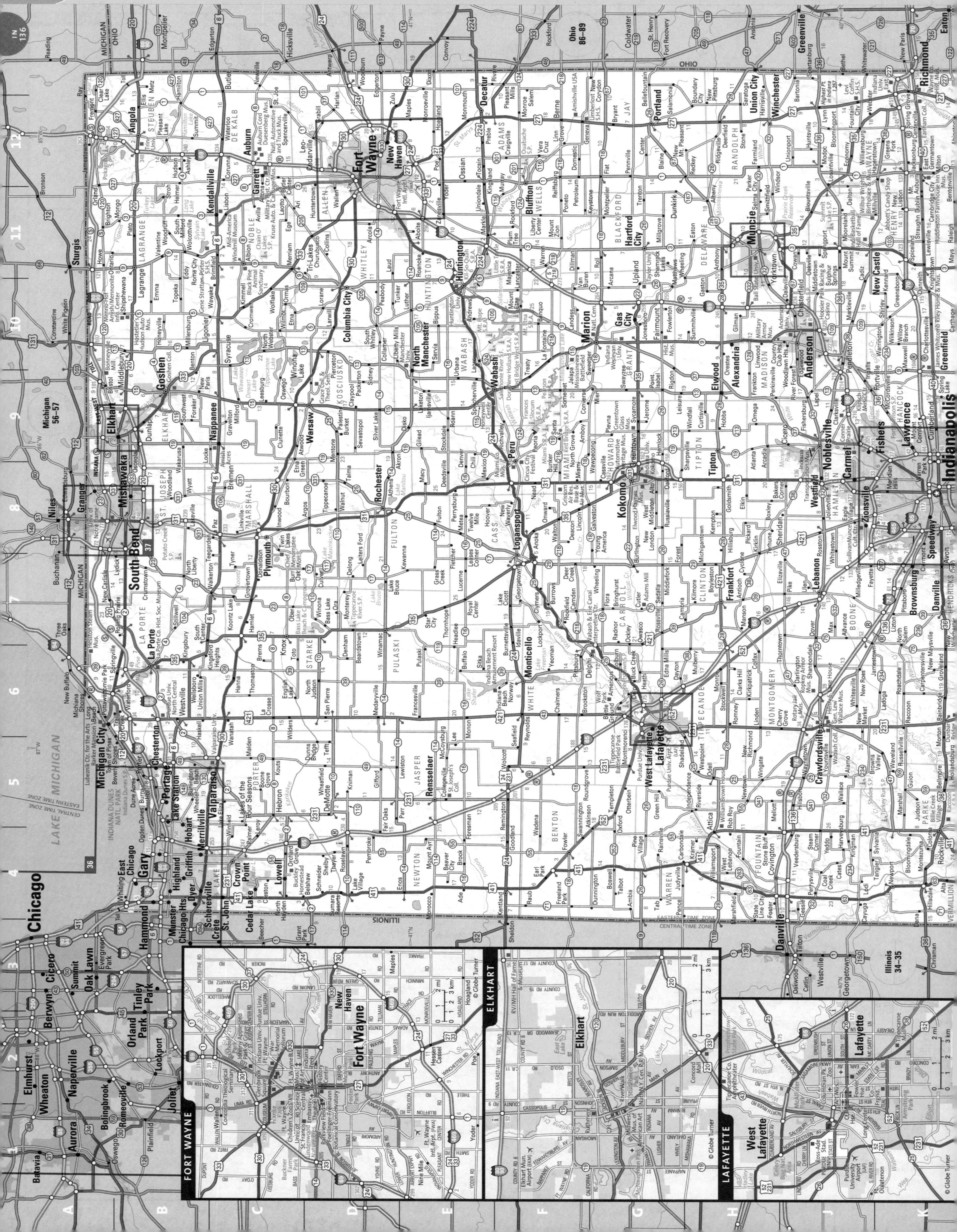

DRIVING DISTANCES IN MILES

SEE ALSO MILEAGE AND DRIVING TIME MAP ON PAGE 144

	ANDERSON	BLOOMINGTON	COLUMBUS	CRAWFORDSVILLE	EVANSVILLE	FORT WAYNE	GARY	GREENSBURG	GREENSBURG	INDIANAPOLIS	KOKOMO	LAFAYETTE	LOUISVILLE, KY.	MUNCIE	PLYMOUTH	RICHMOND	SOUTH BEND	TERRE HAUTE	VINCENNES
EVANSVILLE	211	117	175	164		296	324	195	166	219	194	114	228	281	241	305	107	51
FORT WAYNE	86	175	169	163	296		143	149	128	85	116	236	75	65	95	79	207	253
GARY	180	192	200	122	324	143		210	153	127	91	268	198	66	226	62	161	268
INDIANAPOLIS	43	47	45	47	166	128	153	55		52	66	112	61	74	138	77	123	
SOUTH BEND	129	187	187	136	305	79	62	187	138	86	104	255	141	124	172		217	263

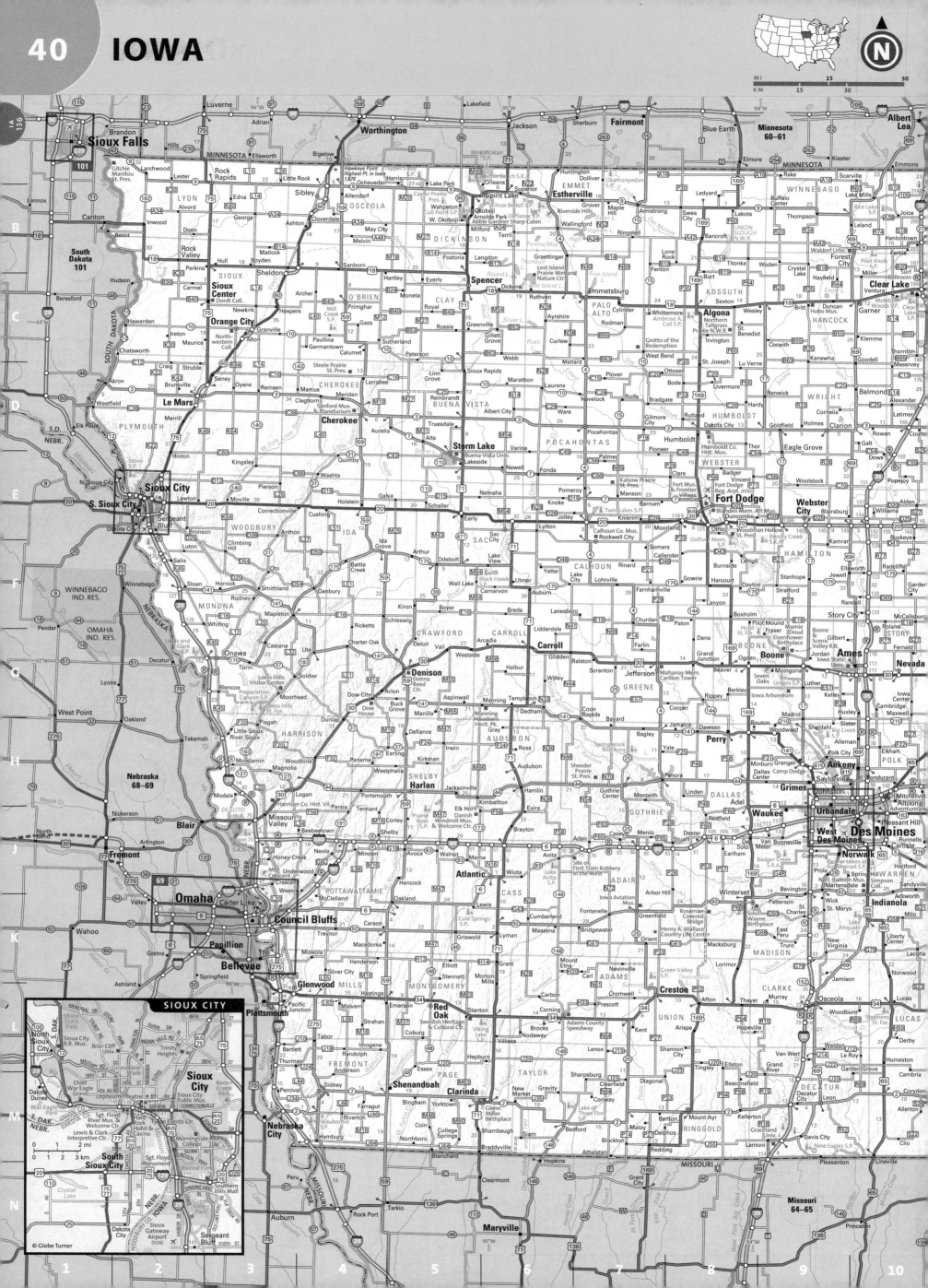

	AMES	BURLINGTON	CARROLL	CEDAR RAPIDS	COUNCIL BLUFFS	CRESTON	DAVENPORT	DECORAH	DES MOINES	DUBUQUE	FORT DODGE	IOWA CITY	MARSHALLTOWN	MASON CITY	OTTUMWA	SIOUX CITY	SPENCER	WATERLOO
COUNCIL BLUFFS	165	323	101	261		99	303	347	130	327	160	245	181	285	216	101	157	238
DES MOINES	34	157	90	129	130	81	171	215		196	94	113	49	126	86	202	188	106
IOWA CITY	136	82	195	28	245	195	59	131	113	84	196		98	157	83	316	267	78
SIOUX CITY	171	394	105	332	101	189	375	303	202	321	120	316	252	218	287		103	228
WATERLOO	95	157	160	53	238	189	137	79	106	93	108	78	58	79	125	228	189	

DRIVING DISTANCES IN MILES   SEE ALSO MILEAGE AND DRIVING TIME MAP ON PAGE 144

DRIVING DISTANCES IN MILES

SEE ALSO MILEAGE AND DRIVING TIME MAP ON PAGE 144

	ATCHISON / ARKANSAS CITY	COLBY	DODGE CITY	EMPORIA	GARDEN CITY	GREAT BEND	HAYS	HUTCHINSON	INDEPENDENCE	IOLA	KANSAS CITY	LAWRENCE	LIBERAL	MANHATTAN	SALINA	TOPEKA	WICHITA
DODGE CITY	141	107	315	238	52	83	106	120	270	264	333	298	83	232	164	271	153
KANSAS CITY	247	50	369	333	106	373	250	261	240	162	105	35	402	117	172	61	192
SALINA	151	160	200	164	118	204	81	93	68	206	187	172	137	247	72	111	92
TOPEKA	193	49	308	271	58	311	188	200	178	135	100	61	26	347	55	111	137
WICHITA	61	186	289	153	85	205	119	181	51	118	192	159	210	131	92	137	

Nebraska 68–69

Oklahoma 90–91

Missouri 64–65

WICHITA (inset map)

© Globe Turner

	ASHLAND	BOWLING GREEN	CINCINNATI, OH	ELIZABETHTOWN	FRANKFORT	GLASGOW	HAZARD	HENDERSON	HOPKINSVILLE	LEXINGTON	LONDON	LOUISVILLE	MAYFIELD	MAYSVILLE	MIDDLESBORO	OWENSBORO	PADUCAH	PIKEVILLE
BOWLING GREEN	274		212	70	161	36	200	107	63	157	145	112	146	222	203	76	135	265
LEXINGTON	119	157	85	89	29	138	120	201	215		77	87	136	183	262	142		
LOUISVILLE	194	112	100	44	54	92	194	123	170	80	156		228	141	214	109	217	217
OWENSBORO	300	76	206	95	161	111	275	30	80	183	221	109	138	248	279		127	323
PADUCAH	379	135	317	175	266	173	337	121	72	262	283	217	24	327	373	127		402

DRIVING DISTANCES IN MILES

SEE ALSO MILEAGE AND DRIVING TIME MAP ON PAGE 144

LEXINGTON

LAND BETWEEN THE LAKES

FRANKFORT

© Globe Turner

	ALEXANDRIA	BATON ROUGE	DE RIDDER	FERRIDAY	HAMMOND	HOUMA	LAFAYETTE	LAKE CHARLES	MONROE	NATCHITOCHES	NEW IBERIA	NEW ORLEANS	OPELOUSAS	RUSTON	SHREVEPORT	SLIDELL	TALLULAH	WINNFIELD
BATON ROUGE	140		176	104	51	101	60	132	188	195	79	91	80	233	261	97	163	188
LAFAYETTE	87	60		119	152	106		76	183	142	20		180	208	152	214	135
MONROE	96	188	170	85	248	283	183	196		100	202	302	156	37	103	298	55	64
NEW ORLEANS	227	91	262	191	57	57	147	219	302	282	166		167	339	348	31	247	275
SHREVEPORT	121	261	137	188	308	309	208	186	103	73	227	348	182		354	158	105

DRIVING DISTANCES IN MILES   SEE ALSO MILEAGE AND DRIVING TIME MAP ON PAGE 144

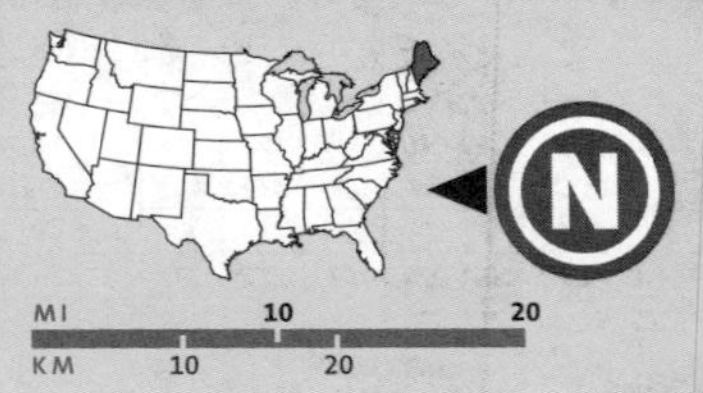

N

MI 10 20
KM 10 20

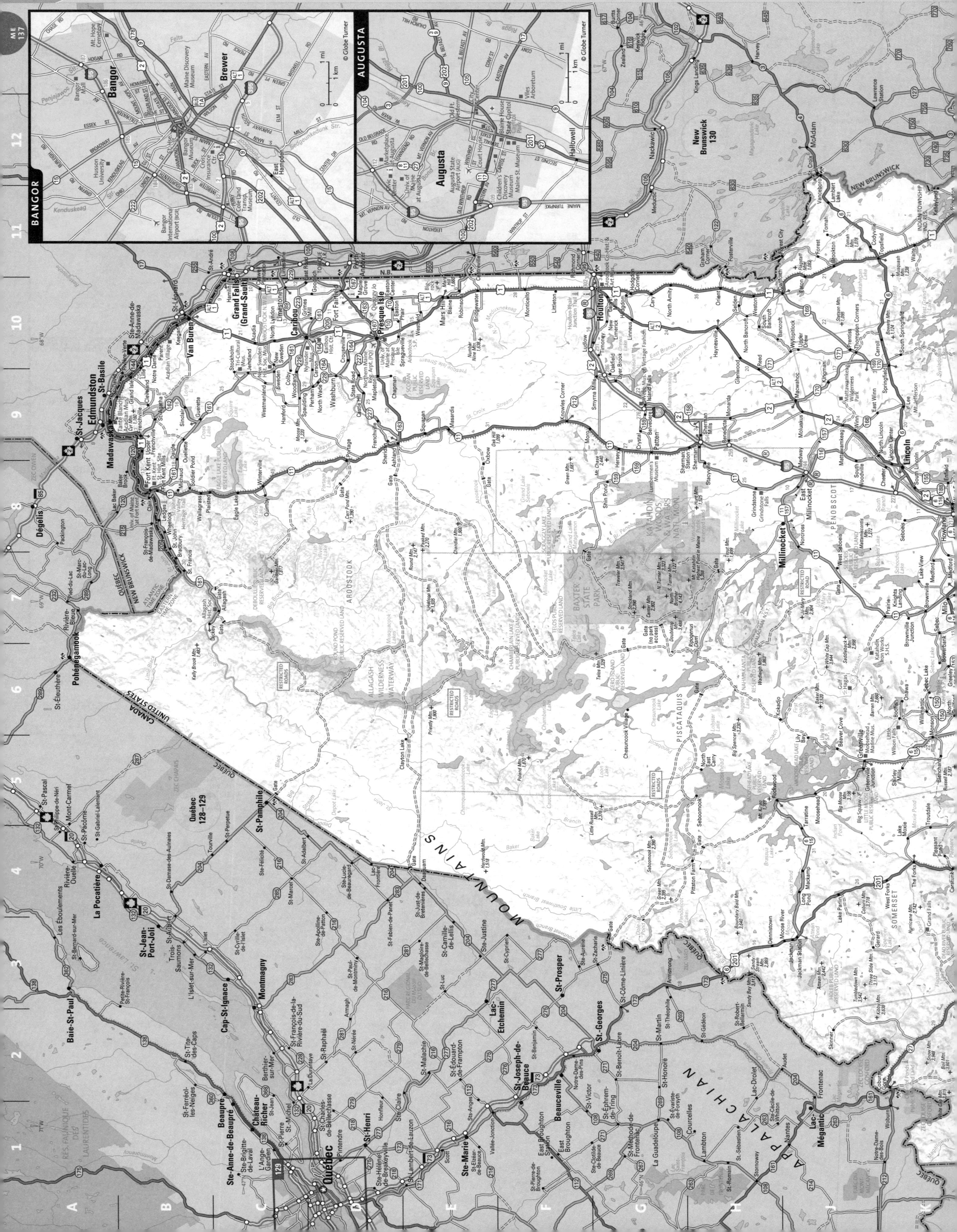

BANGOR

Bangor

Brewer

© Globe Turner

AUGUSTA

Augusta

© Globe Turner

New Brunswick 130

NEW BRUNSWICK

QUEBEC

CANADA
UNITED STATES

ATLANTIC TIME ZONE

Québec 128-129

Québec

AROOSTOOK

PISCATAQUIS

SOMERSET

PENOBSCOT

BAXTER STATE PARK

ALLAGASH WILDERNESS WATERWAY

APPALACHIAN MOUNTAINS

Edmundston
Madawaska
Van Buren
Caribou
Presque Isle
Grand Falls (Grand-Sault)
Houlton
Millinocket
Lincoln
Greenville
St-Pamphile
St-Georges
Beauceville
Ste-Marie
Lac-Mégantic
Montmagny
La Pocatière
Baie-St-Paul
Pohénégamook
Dégelis

	AUGUSTA	BANGOR	BAR HARBOR	BRUNSWICK	CALAIS	FARMINGTON	FORT KENT	GREENVILLE	HOULTON	LEWISTON	MACHIAS	MILLINOCKET	PORTLAND	PORTSMOUTH NH	PRESQUE ISLE	ROCKLAND	SACO	WATERVILLE
AUGUSTA		77	120	32	173	65	269	99	196	35	161	149	58	110	236	43	74	20
BANGOR	77		45	106	97	80	195	74	122	108	85	75	131	184	162	58	147	56
CALAIS	173	97	112	203		177	189	160	91	205	55	112	228	287	133	155	244	153
HOULTON	196	122	166	226	91	200	98	155		228	126	73	251	304	42	182	267	176
PORTLAND	58	131	175	27	228	81	324	153	251	36	216	203		53	291	78	16	84

DRIVING DISTANCES IN MILES

SEE ALSO MILEAGE AND DRIVING TIME MAP ON PAGE 144

MI 10 20
KM 10 20

DOWNTOWN BALTIMORE

BALTIMORE

CUMBERLAND

HAGERSTOWN

© Globe Turner

MARYLAND

51

	ABERDEEN	ANNAPOLIS	BALTIMORE	CAMBRIDGE	CHESTERTOWN	CUMBERLAND	EASTON	FREDERICK	HAGERSTOWN	HANCOCK	LEXINGTON PARK	OCEAN CITY	POCOMOKE CITY	ROCKVILLE	ST. CHARLES	SALISBURY	WASHINGTON DC	WESTMINSTER
ANNAPOLIS	54		25	55	45	162	38	73	98	124	66	108	112	47	47	83	31	56
BALTIMORE	35	25		78	68	140	61	51	76	102	95	131	135	45	57	106	38	39
HAGERSTOWN	109	98	76	153	143	67	136	28		29	142	206	211	54	103	182	70	50
SALISBURY	124	83	106	32	81	246	47	156	182	207	149	30	29	130	130		115	138
WASHINGTON, DC	71	31	38	87	76	134	70	44	70	96	63	139	144	19	25	115		53

DRIVING DISTANCES IN MILES

SEE ALSO MILEAGE AND DRIVING TIME MAP ON PAGE 144

FREDERICK

ANNAPOLIS

© Globe Turner

	BOSTON	BROCKTON	FALL RIVER	FALMOUTH	FITCHBURG	GLOUCESTER	GREENFIELD	HYANNIS	LOWELL	NEW BEDFORD	NORTH ADAMS	NORTHAMPTON	PLYMOUTH	PROVIDENCE	PROVINCETOWN	SPRINGFIELD	WORCESTER
BOSTON		26	53	72	49	35	93	72	31	60	133	105	41	52	117	95	46
NEW BEDFORD	60	38	16	41	101	95	136	45	86		189	137	172	43	33	91	127	78
PITTSFIELD	140	153	153	189	101	174	52	193	142	172	17	41		166	138	239	55	101
SPRINGFIELD	95	108	92	143	85	129	40	148	96	127	72	18	65	120		75	193	55
WORCESTER	46	59	65	95	31	80	58	99	44	78	97	65	101	72	43	144	55	

DRIVING DISTANCES IN MILES   SEE ALSO MILEAGE AND DRIVING TIME MAP ON PAGE 144

NORTHWESTERN MICHIGAN

SAGINAW

ISLE ROYALE NATIONAL PARK

	ALPENA	ANN ARBOR	BENTON HARBOR	CADILLAC	DETROIT	ESCANABA	FLINT	GRAND RAPIDS	HOUGHTON	KALAMAZOO	LANSING	MACKINAW CITY	MARQUETTE	MUSKEGON	PORT HURON	SAGINAW	SAULT STE. MARIE	TRAVERSE CITY
DETROIT	242	42	186	209		438	62	153	556	136	86	291	455	191	58	97	346	257
GRAND RAPIDS	261	129	78	99	153	391	112		510	53	67	244	408	40	176	144	299	141
LANSING	230	63	126	131	86	375	53	67	493	76		228	391	105	117	86	282	173
MACKINAW CITY	94	281	323	145	291	149	230	244	268	302	228		166	248	293	198	57	106
MARQUETTE	257	444	487	309	455	65	393	408	102	466	391	166		412	457	361	163	269

DRIVING DISTANCES IN MILES   SEE ALSO MILEAGE AND DRIVING TIME MAP ON PAGE 144

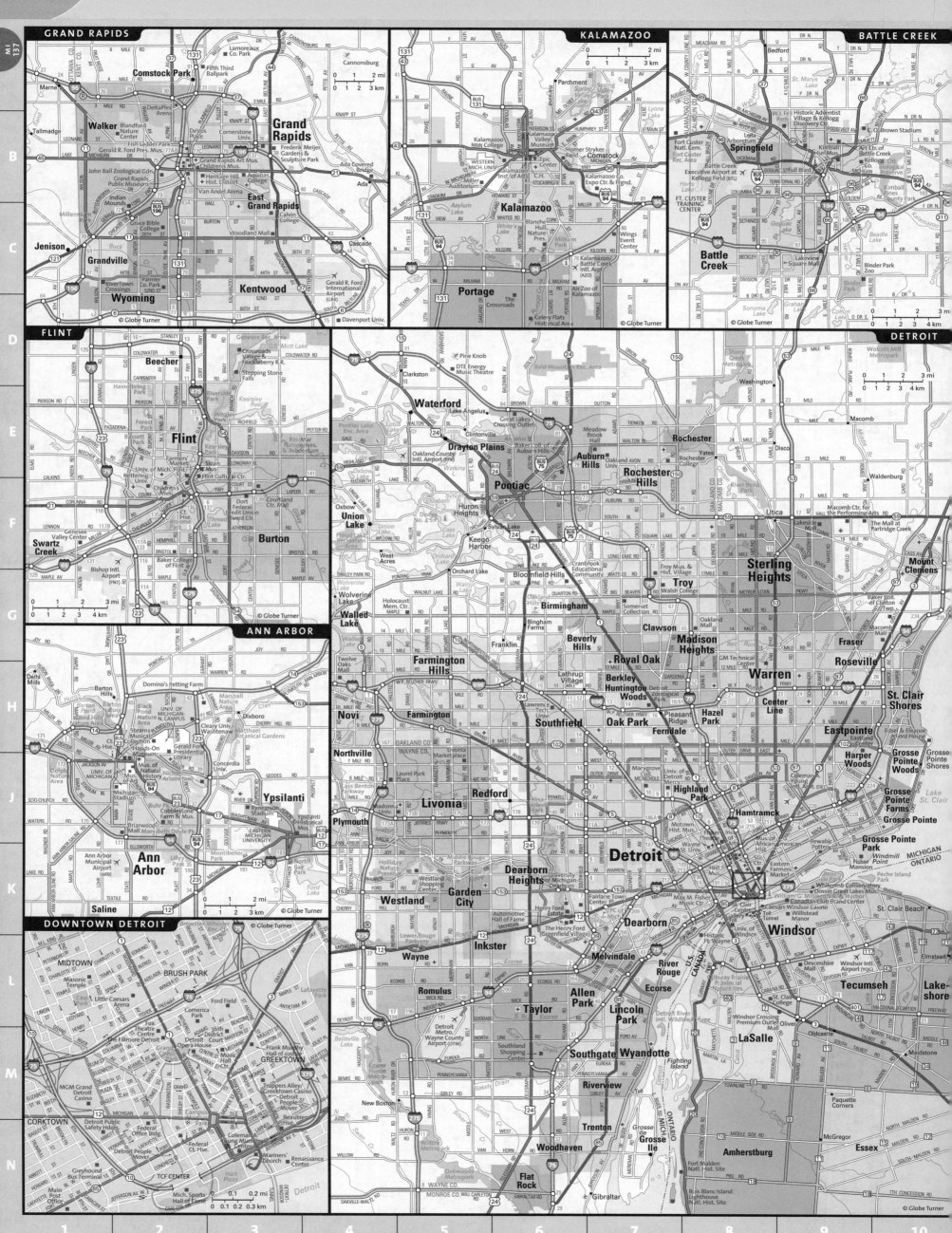

DOWNTOWN MINNEAPOLIS

DOWNTOWN ST. PAUL

© Globe Turner

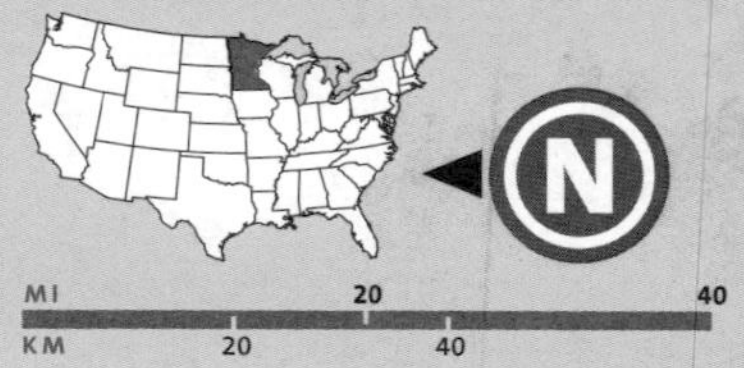

MI 20 40
KM 20 40

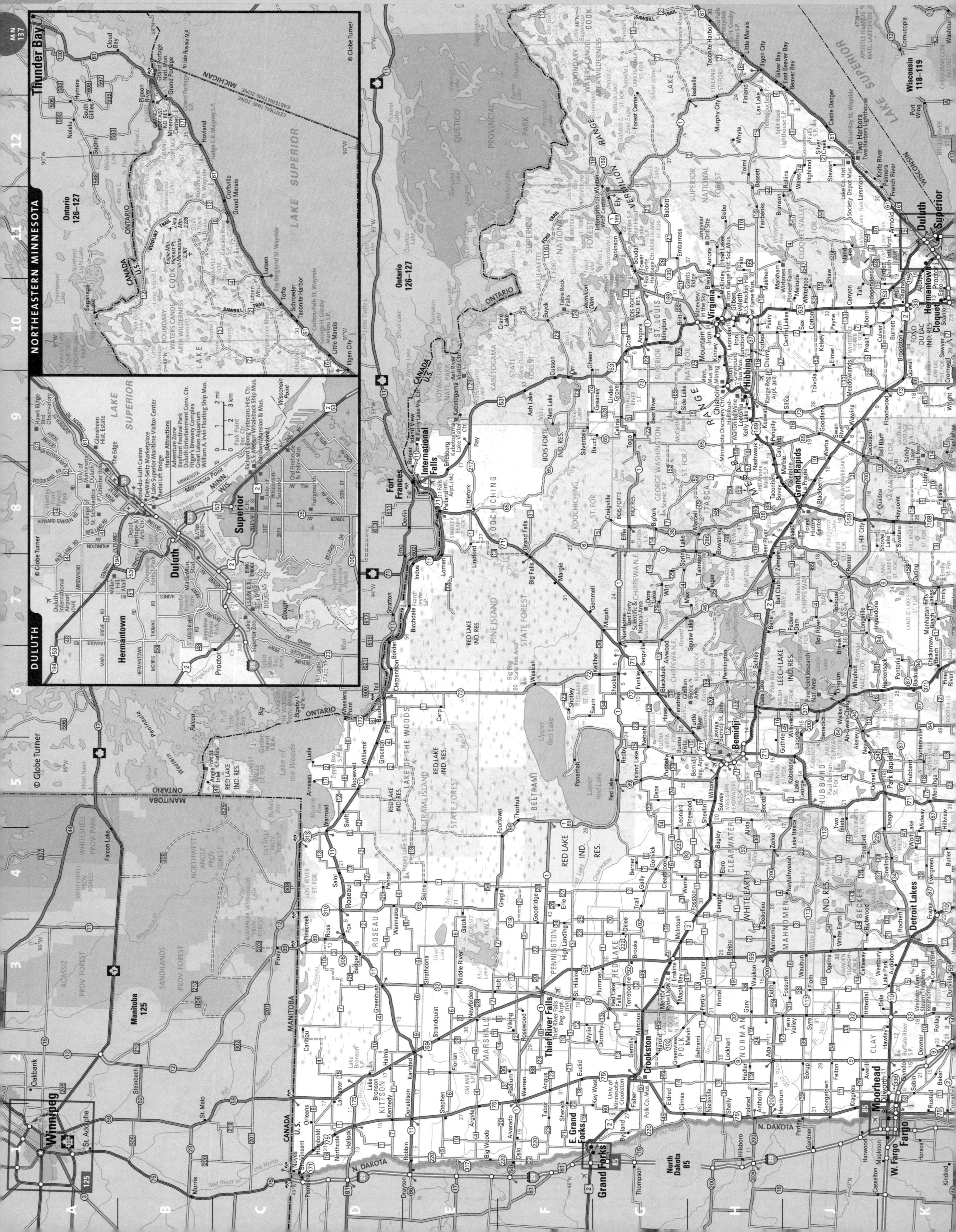

NORTHEASTERN MINNESOTA

Ontario 126–127

Thunder Bay

DULUTH

Wisconsin 118–119

Ontario 126–127

Manitoba 125

LAKE SUPERIOR

International Falls

Fort Frances

Bemidji

Grand Rapids

Hibbing

Virginia

Duluth
Superior

Detroit Lakes

Thief River Falls

Crookston

E. Grand Forks
Grand Forks

Moorhead
Fargo
W. Fargo

Winnipeg

N. Dakota 85

North Dakota 85

© Globe Turner

	DULUTH	MINNEAPOLIS	MOORHEAD	ROCHESTER	ST. CLOUD
BEMIDJI / ALBERT LEA	251	96	325	62	159
BRAINERD	153	225	133	311	157
DULUTH	116	129	141	215	62
FERGUS FALLS		158		239	149
GRAND FORKS, ND.	210	176	257	262	117
INTERNATIONAL FALLS	264	312	55	399	254
MANKATO	157	290	80	371	251
MARSHALL	239	78	242	80	128
MINNEAPOLIS	274	148	294	185	131
MOORHEAD	158		220	88	64
ROCHESTER	257	230	230		72
ST. CLOUD	239	88	317	317	151
ST. PAUL	149	64	172	151	
VIRGINIA	154	10	29	80	80
WILLMAR	61	193	229	274	274
WINONA	206	92	168	177	63
WORTHINGTON	216	120	351	51	185
	362	207	298	174	201

DRIVING DISTANCES IN MILES   SEE ALSO MILEAGE AND DRIVING TIME MAP ON PAGE 144

ROCHESTER

© Globe Turner

	BILOXI	COLUMBUS	GREENVILLE	HATTIESBURG	JACKSON	MEMPHIS TN	MERIDIAN	NATCHEZ	NEW ORLEANS LA	TUPELO	VICKSBURG	WINONA
BILOXI		262	297	82	172	379	171	231	93	317	214	262
GREENVILLE	297	164		216	125	148	216	157	102	191	89	82
JACKSON	172	153	125	90		211	91	107	185	175	42	94
MERIDIAN	171	91	216	89	91	234		194	201	146	133	113
TUPELO	317	66	172	235	175	109	146	269	347		213	99

DRIVING DISTANCES IN MILES

SEE ALSO MILEAGE AND DRIVING TIME MAP ON PAGE 144

MI 25 50
KM 25 50

N

Inset maps: JACKSON, HATTIESBURG, VICKSBURG, BILOXI–GULFPORT

© Globe Turner

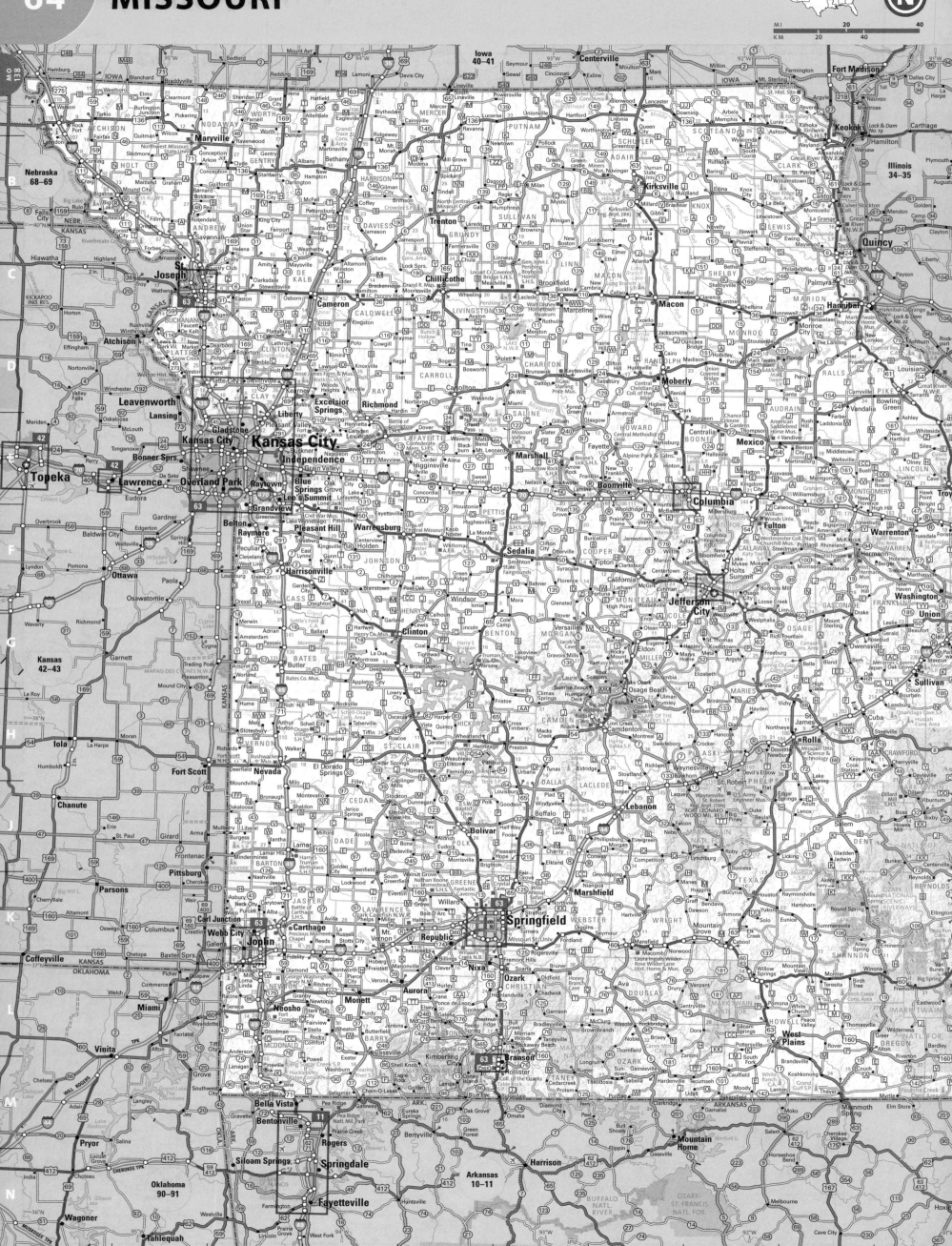

	BRANSON	CAPE GIRARDEAU	CHILLICOTHE	COLUMBIA	HANNIBAL	JEFFERSON CITY	JOPLIN	KANSAS CITY	KIRKSVILLE	NEVADA	POPLAR BLUFF	ROLLA	ST. JOSEPH	ST. LOUIS	SEDALIA	SIKESTON	SPRINGFIELD	WEST PLAINS
CAPE GIRARDEAU	347		357	234	228	243	382	363	322	374	75	205	419	120	303	36	307	175
COLUMBIA	203	234	124		101	32	238	129	89	206	269	97	185	123	69	265	163	194
KANSAS CITY	210	363	92	129	230	161	165		161	104	398	226	56	252	97	394	169	278
ST. LOUIS	249	120	247	123	117	132	284	252	212	276	156	107	308		192	151	209	204
SPRINGFIELD	41	307	200	163	241	131	77	169	251	95	191	110	225	209	108	238		109

DRIVING DISTANCES IN MILES

SEE ALSO MILEAGE AND DRIVING TIME MAP ON PAGE 144

ST. LOUIS

PORTAGE DES SIOUX

DOWNTOWN ST. LOUIS

JEFFERSON CITY

COLUMBIA

Illinois 34–35

Arkansas 10–11

Kentucky 44–45

Tennessee 102–103

© Globe Turner

MI 25 50
KM 25 50

N

BRITISH COLUMBIA
Brit. Col. 122
Alberta 123
ALBERTA
CANADA
UNITED STATES

Roosville
Eureka
Bonners Ferry
Libby
Troy
Kalispell
Whitefish
Evergreen
Columbia Falls
Polson
Pablo
Ronan
St. Ignatius
Thompson Falls
Plains
Kellogg
Wallace
St. Regis
Superior
Missoula
Lolo
Stevensville
Hamilton
Darby
Orofino
Kamiah
Kooskia

Cardston
WATERTON
GLACIER NATIONAL PARK
BLACKFEET IND. RES.
Browning
St. Mary
Cut Bank
Shelby
Conrad
Choteau
Great Falls
Augusta
Helena
East Helena
Deer Lodge
Anaconda
Butte
Whitehall
Boulder
Townsend
Three Forks
Belgrade
Bozeman
Livingston
Dillon
Lima
West Yellowstone
YELLOWSTONE NATL. PARK

Fort Benton
Shonkin
CHOUTEAU
LIBERTY
HILL
TOOLE
PONDERA
TETON
CASCADE
LEWIS AND CLARK
JUDITH BASIN
MEAGHER
BROADWATER
JEFFERSON
GALLATIN
MADISON
BEAVERHEAD
DEER LODGE
POWELL
MISSOULA
RAVALLI
GRANITE
SILVER BOW

KOOTENAI NATL. FOR.
FLATHEAD NATIONAL FOREST
LOLO NATL. FOR.
BITTERROOT NATIONAL FOREST
BEAVERHEAD DEERLODGE NATL. FOR.
HELENA NATL. FOR.
LEWIS AND CLARK NATL. FOR.
GALLATIN NATL. FOREST
CARIBOU-TARGHEE NATL. FOR.
SALMON-CHALLIS NATIONAL FOREST

ROCKY MOUNTAINS
MISSION RANGE
SWAN RANGE
BITTERROOT RANGE
SAPPHIRE MTS.
ANACONDA RANGE
PIONEER MTS.
BEAVERHEAD MTS.
TOBACCO ROOT MTS.
GALLATIN RANGE
CRAZY MTS.
ABSAROKA RANGE
GRAVELLY RANGE
MADISON RANGE
BIG BELT MOUNTAINS
LITTLE BELT MOUNTAINS

IDAHO
Idaho 33
WYOMING
Wyoming 120

MISSOULA

GLACIER NATIONAL PARK
Cardston
Kalispell
Evergreen
Whitefish
Columbia Falls

Idaho Falls
Rexburg
Rigby
St. Anthony
Jackson
GRAND TETON NATIONAL PARK
TETON RANGE

© Globe Turner

DRIVING DISTANCES IN MILES

SEE ALSO MILEAGE AND DRIVING TIME MAP ON PAGE 144

	Billings	Bozeman	Browning	Butte	Dillon	Glasgow	Glendive	Great Falls	Havre	Helena	Kalispell	Lewistown	Miles City	Missoula	Shelby	Sheridan, Wy	West Glacier	West Yellowstone
Billings		141	346	223	253	277	217	222	249	235	454	125	144	340	304	131	474	232
Butte	223	81	241		63	430	439	153	354	71	232	247	367	118	235	354	252	162
Great Falls	222	176	124	153	215	277	351		118	85	222	109	329	199	82	353	192	257
Helena	235	94	174	71	132	362	452	85	203		195	192	379	114	168	366	216	174
Missoula	340	199	201	118	171	476	557	199	317	114	116	306	484		227	471	136	279

MI 20 40
KM 20 40

NORTH PLATTE

Homestead Rd
Lincoln Co. Frgrds.
SCOUTS REST
Buffalo Bill Ranch S.H.P.
Ranch Rd
Wild West Arena
Lincoln Co. Hist. Museum
Golden Spike Tower
Cody Park
North Platte River's Edge
City Hall
Rec. Complex
W South River Rd
Platte River Mall
North Platte Community Coll. North Campus
Indian Meadows
W State Farm Rd
North Platte Community Coll. South Campus
E State Farm Rd
© Globe Turner

KEARNEY

Kearney Area Children's Mus.
Hilltop Mall
Mus. of Nebraska Art
Buffalo Co. Hist. Museum
Univ. of NE-Kearney
Merryman Perf. Arts Ctr.
Yanney Heritage Park
Classic Car Collection
Kearney
C.H.
Fort Kearny Mus.
Nebraska Firefighters Mus.
Younes Conference Ctr.
© Globe Turner

GRAND ISLAND

Central Nebraska Regional Airport (GRI)
Eagle Scout Park
Conestoga Mall
Burlington Station
Court House
Stuhr Museum of the Prairie Pioneer
Island Oasis Water Park & Nebr. State Frgrds.
HeartLand Events Ctr.
Fonner Park Race Track
© Globe Turner

Rapid City
Newcastle
Hill City
Mt. Rushmore Natl. Mem.
BLACK HILLS NATL. FOR.
Custer
Jewel Cave Natl. Mon.
CUSTER S.P.
WIND CAVE NATL. PARK
Hot Springs
Edgemont
Ardmore
BUFFALO GAP NATL. GRASSLAND
BADLANDS NATL. PARK
Porcupine
Oelrichs
PINE RIDGE INDIAN RESERVATION
Whiteclay
SOUTH DAKOTA
Wounded Knee
Martin
Tuthill
St. Francis
Olsonville
ROSEBUD INDIAN RES.
Rosebud
South Dakota 101

Wyoming
Newcastle
Torrington
Van Tassell
Harrison
Crawford
FORT ROBINSON
Fort Robinson Museum
Chadron
Museum of the Fur Trade
Chadron State Coll.
Whitney
Marsland
Agate Fossil Beds Natl. Mon.
SIOUX
Hemingford
BOX BUTTE
Berea
Alliance
Knight Mus. of High Plains Heritage
Alliance Mun. Arpt. (AIA)
Antioch
Lakeside
Ellsworth
Bingham
Ashby
Whitman
Hyannis
Mullen
Seneca
Thedford
Halsey
HOOKER
GRANT
THOMAS
BLAINE
Dunning
Milburn
Brewster
Merna
CUSTER
Gordon
Clinton
Hay Springs
Rushville
Sheridan Co. Hist. Mus.
Tri-State Old Time Cowboys Mem. Mus.
SHERIDAN
Mari Sandoz St. Hist. Marker
Big Hill 4,144
SURVEY VALLEY
SAND HILLS
Cody
Nenzel
Kilgore
Crookston
Sandhills Hall
Valentine
FORT NIOBRARA N.W.R.
Sparks
Norden
Springview
Niobrara Valley Preserve
Keller Park St. Rec. Area
Wood Lake
CHERRY
SAMUEL R. McKELVIE NATL. FOR.
Merritt Res. St. Rec. Area
VALENTINE N.W.R.
Ainsworth
Johnstown
Long Pine
BROWN
Elsmere
Purdum
Giant Hill 3,400
Brownlee
Scott Lookout Tower
Halsey
NEBRASKA NATL. FOR.

Wild Horse Hill 4,204
Alliance
North Platte N.W.R.
Scottsbluff
Gering
Mitchell
Morrill
Lyman
Henry
Scotts Bluff Natl. Mon.
Riverside Discovery Ctr.
North Platte Valley Mus.
SCOTTS BLUFF
Melbeta
Minatare
Angora
Bayard
Chimney Rock N.H.S.
Bridgeport
Northport
McGrew
Broadwater
Lisco
Oshkosh
Blue Water Battlefield
GARDEN
Courthouse Mus. & Baled Hay Church
Lake McConaughy St. Rec. Area
Arthur
ARTHUR
McPHERSON
Tryon
Ringgold
Stapleton
Gandy
Arnold
LOGAN
Anselmo
Merna

La Grange
Harrisburg
BANNER
Redington
Courthouse Rock and Jail Rock
Dalton
Gurley
MORRILL
CHEYENNE
Kimball
Bushnell
Dix
Potter
Brownson
Fort Sidney Mus. and Post Commander's Home
Sidney
Sunol
Lodgepole
Colton
KIMBALL
Panorama Point Highest Point in Nebraska 5,424
COLORADO
Pine Bluffs
Chappell
DEUEL
Big Springs
Brule
Ogallala
Lake C.W. McConaughy
Kingsley Dam
Lake Ogallala
Keystone
Roscoe
Paxton
Sarben
Sutherland
Hershey
Maxwell
North Platte
Bailey R.R. Yard & Golden Spike Tower
North Platte Reg. Arpt. (LBF)
Brady
Gothenburg
Robert Henri Mus.
Willow Island
Cozad
DAWSON
KEITH
LINCOLN
Fort McPherson Natl. Cem.
Pony Express Sta.
Jeffrey Res.

Peetz
Sedgwick
Julesburg
Ovid
Crook
Proctor
Iliff
Sterling
Stoneham
Atwood
Fleming
Haxtun
Holyoke
PERKINS
Grant
Madrid
Elsie
Wallace
Dickens
Wellfleet
Dancing Leaf Cult. Learning Ctr.
Maywood
Curtis
Moorefield
Farnam
Gallagher Canyon S.R.A.
Eustis
FRONTIER
Stockville
GOSPER
Smithfield
Elwood
Oconto
Callaway
Brandon
Venango
Amherst
Lamar
CHASE
Imperial
Enders
Enders Res.
HAYES
Hayes Center
Palisade
Hamlet
Wauneta
Champion
Red Willow St. Rec. Area
McCook
Ben Nelson Reg. Arpt. (MCK)
Indianola
Cambridge
FURNAS
Arapahoe
Edison
Fort Morgan
Brush
Log Lane Village
Akron
Yuma
Wray
Laird
Parks
Benkelman
Haigler
Max
Stratton
Trenton
Massacre Canyon Mon.
Swanson Res.
Culbertson
Bartley
McCook
Museum of the High Plains
RED WILLOW
HITCHCOCK
DUNDY
Rock Creek Lake St. Rec. Area
Marion
Danbury
Lebanon
Wilsonville
Hendley
Beaver City
Precept
Cedar Bluffs
KANSAS
Kansas 42–43
Norton

WYOMING
SOUTH DAKOTA
COLORADO
KANSAS

MOUNTAIN TIME ZONE
CENTRAL TIME ZONE

© Globe Turner

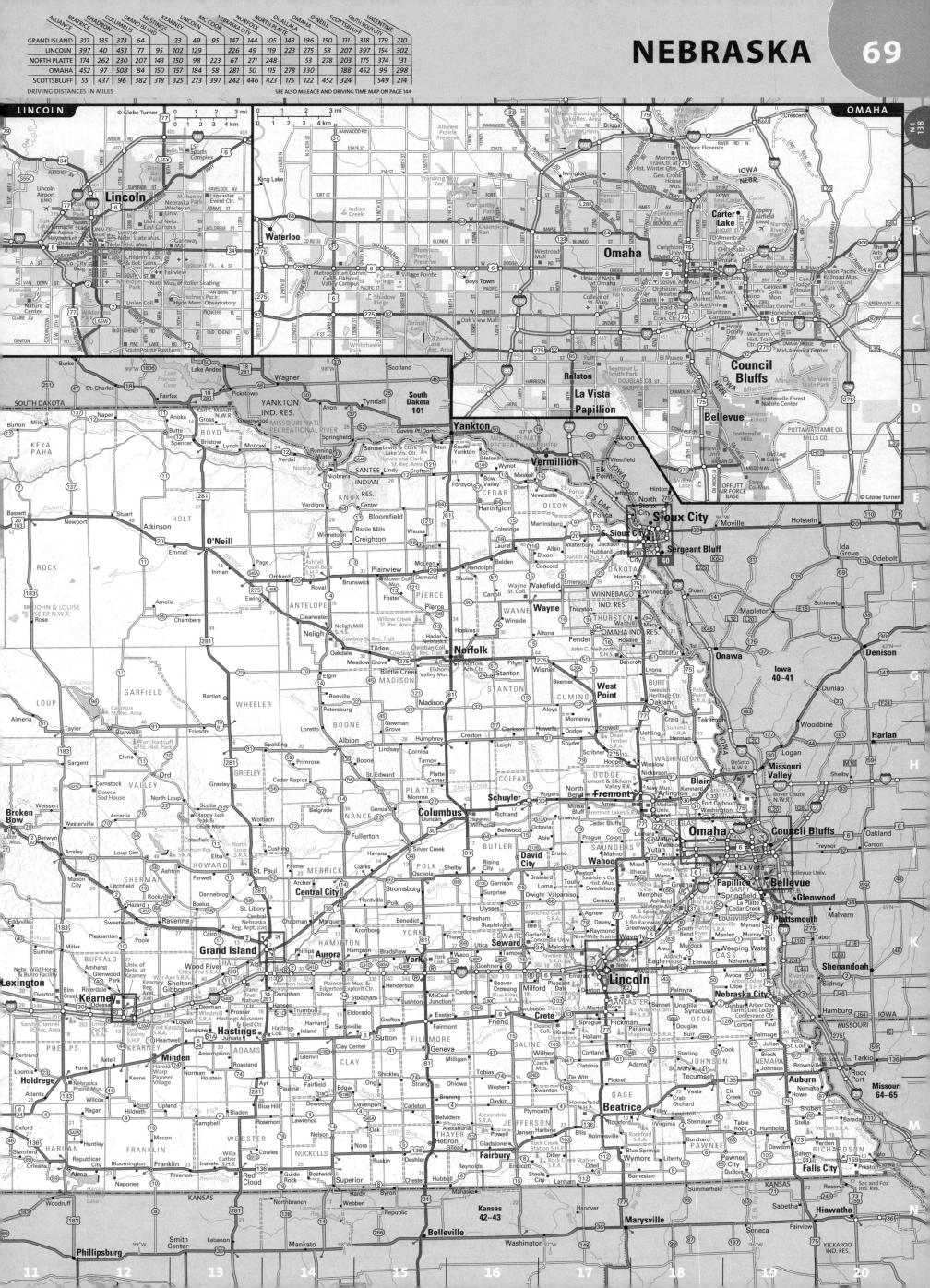

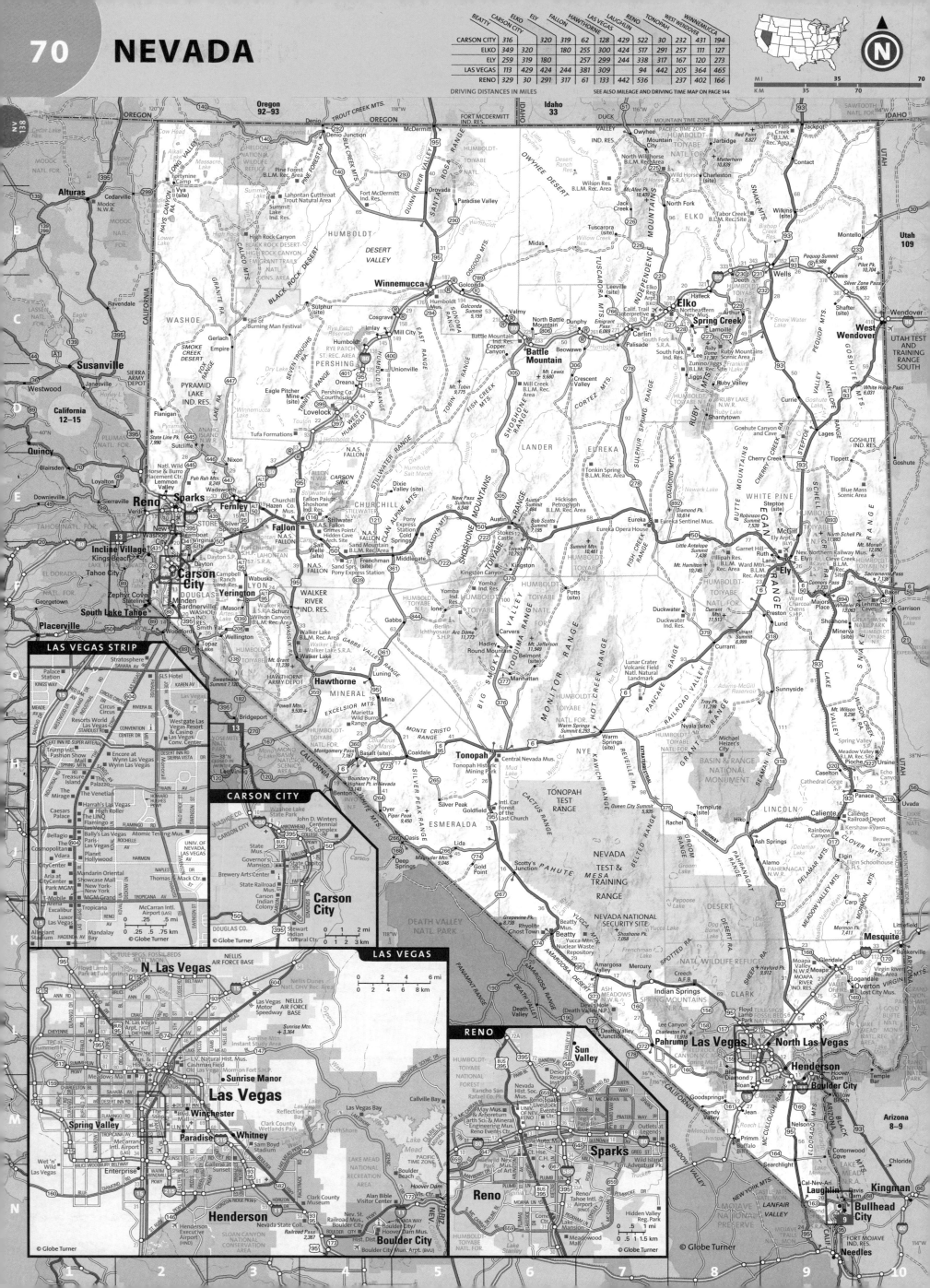

	BEATTY	CARSON CITY	ELKO	ELY	FALLON	HAWTHORNE	LAS VEGAS	LAUGHLIN	RENO	TONOPAH	WEST WENDOVER	WINNEMUCCA
CARSON CITY	316		320	319	62	128	429	522	30	232	431	194
ELKO	349	320		180	255	300	424	517	291	257	111	127
ELY	259	319	180		257	299	244	338	317	167	120	273
LAS VEGAS	113	429	424	244	381	309		94	442	205	364	465
RENO	329	30	291	317	61	133	442	536		237	402	166

DRIVING DISTANCES IN MILES   SEE ALSO MILEAGE AND DRIVING TIME MAP ON PAGE 144

	BERLIN	CONCORD	CONWAY	KEENE	LACONIA	LEBANON	LITTLETON	MANCHESTER	NASHUA	PLYMOUTH	PORTSMOUTH	ROCHESTER
BERLIN		112	42	161	100	114	42	130	144	72	119	96
CONCORD	112		74	50	26	56	84	19	33	41	47	35
LEBANON	114	56	87	66	56		72	71	85	41	102	90
MANCHESTER	130	19	92	51	44	71	102		16	59	45	46
PORTSMOUTH	119	47	77	96	59	102	129	45	56	86		22

DRIVING DISTANCES IN MILES

SEE ALSO MILEAGE AND DRIVING TIME MAP ON PAGE 144

Québec 128–129

Maine 48–49

Maine 54–55

N
MI 10 20
KM 10 20

© Globe Turner

DRIVING DISTANCES IN MILES

SEE ALSO MILEAGE AND DRIVING TIME MAP ON PAGE 144

	Alamogordo	Albuquerque	Carlsbad	Clovis	El Paso, TX	Farmington	Gallup	Hobbs	Las Cruces	Las Vegas	Los Alamos	Raton	Roswell	Santa Fe	Silver City	Socorro	Taos	Tucumcari
ALBUQUERQUE	213		275	220	263	181	141	316	220	115	92	221	199	55	234	77	123	174
FARMINGTON	399	181	455	401	450		120	496	407	264	196	360	379	205	361	263	211	355
LAS CRUCES	65	220	203	293	42	407	338	250		335	312	441	182	275	146	343	394
ROSWELL	117	199	76	110	203	379	340	117	182	178	228	284		191	293	164	248	161
SANTA FE	220	55	267	213	319	205	197	308	275	65	37	171	191		290	132	68	167

Albuquerque (inset map)

Santa Fe (inset map)

Roswell (inset map)

Las Cruces (inset map)

El Paso · Ciudad Juárez

© Globe Turner

NM 138

	ALBANY	BINGHAMTON	BUFFALO	COOPERSTOWN	GENEVA	ITHACA	JAMESTOWN	KINGSTON	LAKE PLACID	NEW YORK	NIAGARA FALLS	OLEAN	PLATTSBURGH	ROCHESTER	SARATOGA SPRINGS	SYRACUSE	UTICA	WATERTOWN
ALBANY		135	292	89	197	186	361	56	138	151	306	297	160	228	32	146	94	179
BINGHAMTON	135		225	78	97	53	214	125	253	176	239	164	276	161	160	76	95	139
BUFFALO	292	225		239	103	153	74	346	400	20	74	374	74	293	152	199	210	
ROCHESTER	228	161	74	175	39	89	142	282	273	336	88	113	310		229	88	135	165
SYRACUSE	146	76	152	93	56	59	220	201	192	250	163	228	88	147		53	65	

DRIVING DISTANCES IN MILES   SEE ALSO MILEAGE AND DRIVING TIME MAP ON PAGE 144

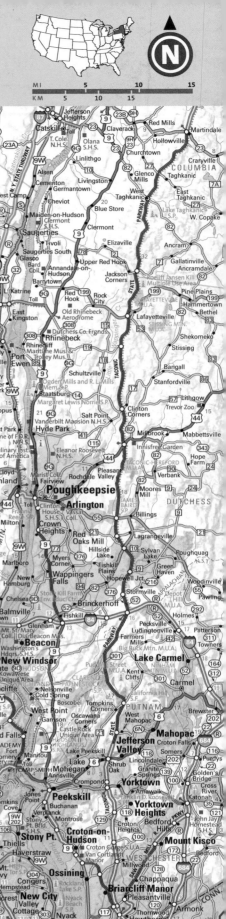

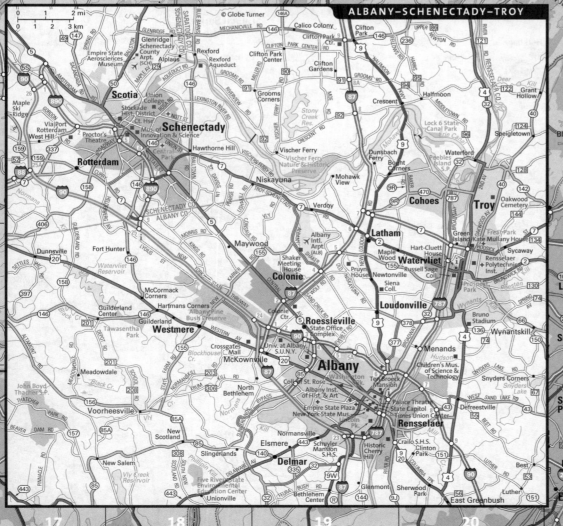

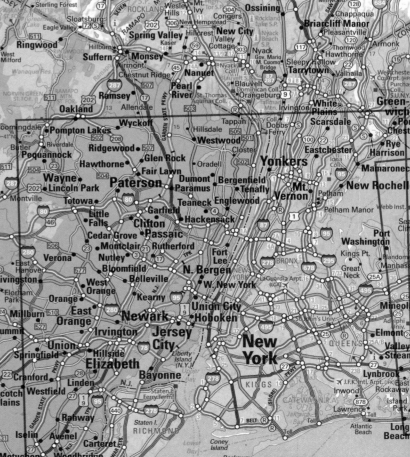

	ALBANY	BINGHAMTON	BUFFALO	HEMPSTEAD	KINGSTON	MIDDLETOWN	MONTAUK	MONTICELLO	NEWBURGH	NEW YORK	PEEKSKILL	PORT JEFFERSON	PORT JERVIS	POUGHKEEPSIE	RIVERHEAD	ROCHESTER	SYRACUSE	WHITE PLAINS
ALBANY		135	292	182	56	105	276	103	89	151	107	214	151	79	234	228	146	148
BINGHAMTON	135		225	203	125	116	296	87	134	176	144	235	117	127	254	161	76	167
KINGSTON	56	125	346	132		55	225	53	39	101	57	164	71	20	183	282	201	98
NEWBURGH	89	134	357	74	39	26	167	47		56	18	106	42	25	125	294	208	42
NEW YORK	151	176	400	26	101	72	120	95	56		45	58	87	96	78	336	250	29

DRIVING DISTANCES IN MILES

SEE ALSO MILEAGE AND DRIVING TIME MAP ON PAGE 144

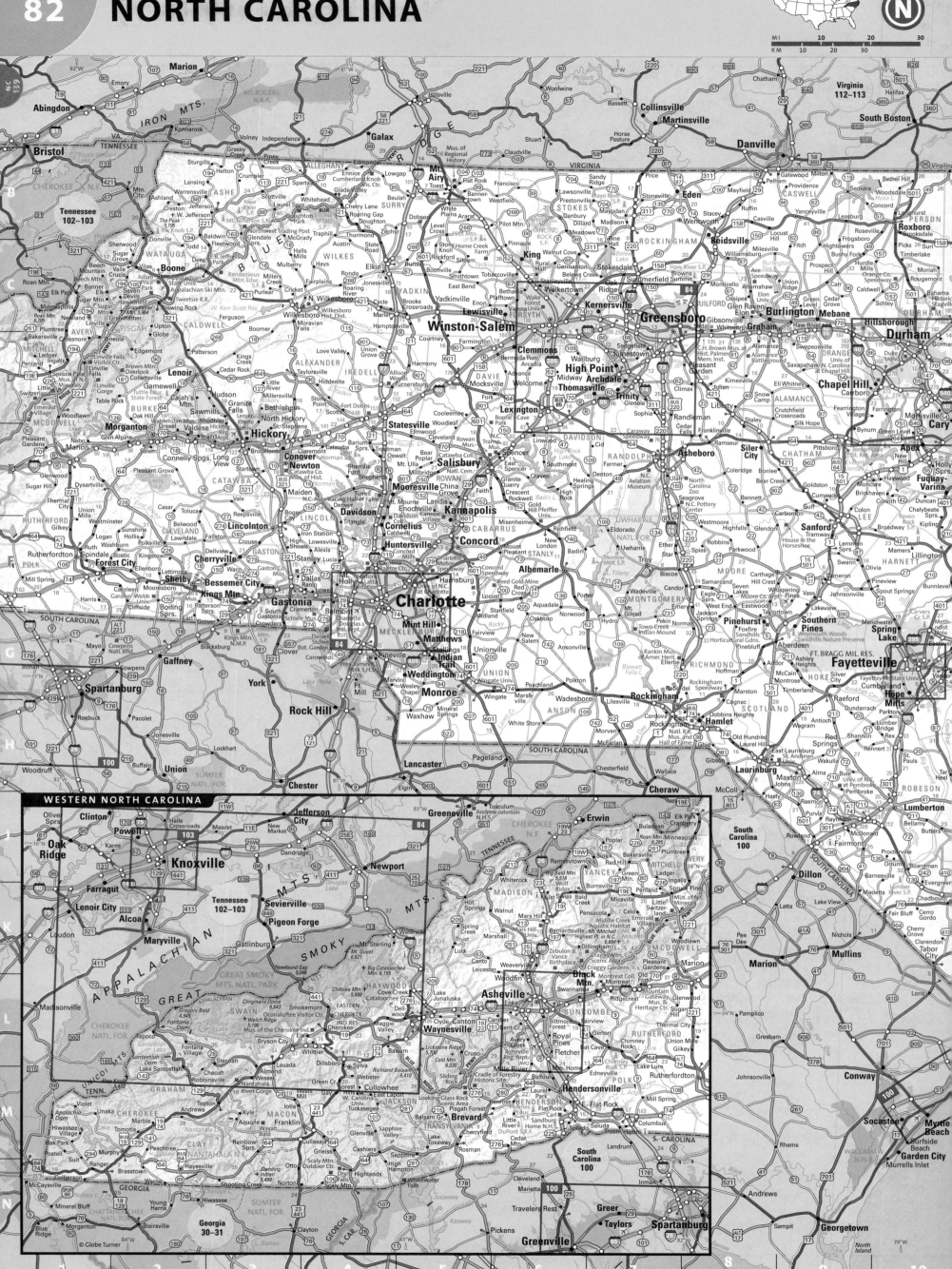

	ASHEVILLE	BOONE	CHARLOTTE	DURHAM	ELIZABETH CITY	FAYETTEVILLE	GREENSBORO	GREENVILLE	HICKORY	JACKSONVILLE	KINSTON	MOREHEAD CITY	NAGS HEAD	RALEIGH	ROCKINGHAM	ROCKY MOUNT	WILMINGTON	WINSTON-SALEM
ASHEVILLE		198	116	224	404	264	176	324	78	354	316	383	444	242	190	297	368	146
CHARLOTTE	116	95		139	319	139	91	239	48	269	232	298	359	158	74	213	205	79
GREENSBORO	176	117	91		49	228	90		148	98	179	141	207	268	67	83	122	193	30
RALEIGH	242	183	158	24	160	62	67	80	164	113	76	142	200		96	54	127	96
WILMINGTON	368	309	205	150	211	92	193	123	290	52	93	95	241	127	131	153		222

DRIVING DISTANCES IN MILES

SEE ALSO MILEAGE AND DRIVING TIME MAP ON PAGE 144

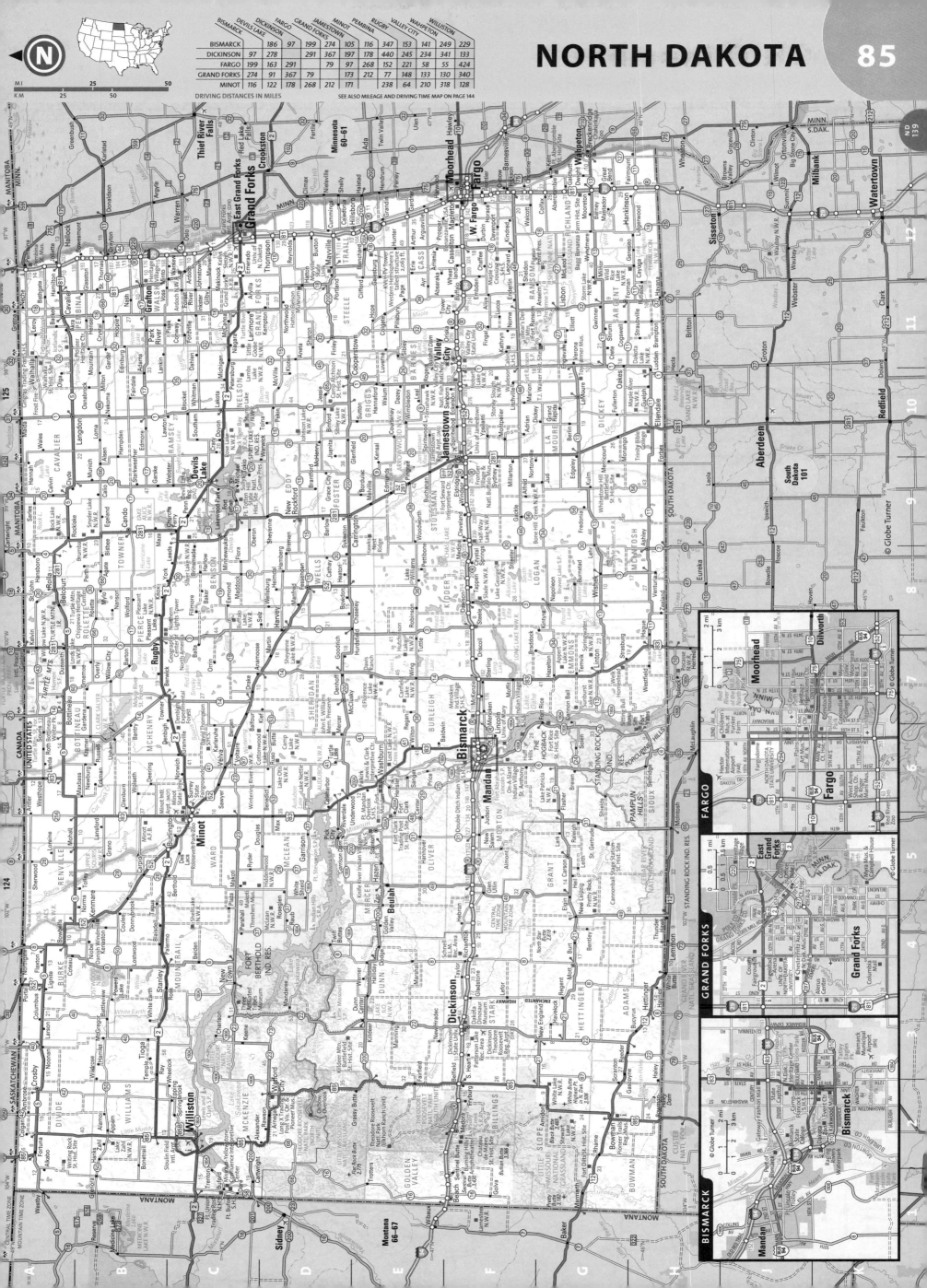

ND
139

DRIVING DISTANCES IN MILES

SEE ALSO MILEAGE AND DRIVING TIME MAP ON PAGE 144

	BISMARCK	DEVILS LAKE	DICKINSON	FARGO	GRAND FORKS	JAMESTOWN	MINOT	PEMBINA	RUGBY	VALLEY CITY	WAHPETON	WILLISTON
BISMARCK		186	97	199	274	105	116	347	153	141	249	229
DICKINSON	97	278		291	367	197	178	440	245	234	341	133
FARGO	199	163	291		79	97	268	152	221	58	55	424
GRAND FORKS	274	91	367	79		173	212	77	148	133	130	340
MINOT	116	122	178	268	212	171		238	64	210	318	128

DRIVING DISTANCES IN MILES

	AKRON	CAMBRIDGE	CANTON	CHILLICOTHE	CINCINNATI	CLEVELAND	COLUMBUS	DAYTON	DEFIANCE	FINDLAY	LIMA	MANSFIELD	MARION	NEW PHILADELPHIA	SANDUSKY	SPRINGFIELD	TOLEDO	YOUNGSTOWN
AKRON		83	23	184	243	38	129	198	186	140	157	66	101	46	84	172	142	49
CAMBRIDGE	83		61	98	187	124	80	155	228	178	175	108	126	37	169	128	228	109
CLEVELAND	38	124	64	199	259		144	213	163	126	163	81	116	87	61	187	119	75
COLUMBUS	129	80	143	47	110	144		70	146	101	96	67	50	117	119	44	148	175
TOLEDO	142	228	168	189	209	119	148	156	63	51	83	105	100	191	62	169		179

SEE ALSO MILEAGE AND DRIVING TIME MAP ON PAGE 144

CLEVELAND

DOWNTOWN CLEVELAND

© Globe Turner

LAKE ERIE

LAKE ERIE

© Globe Turner

	AKRON	ATHENS	CAMBRIDGE	CHILLICOTHE	CINCINNATI	CLEVELAND	COLUMBUS	DAYTON	GALLIPOLIS	HILLSBORO	HUNTINGTON WV	LANCASTER	MARIETTA	PORTSMOUTH	SPRINGFIELD	TOLEDO	WHEELING WV	ZANESVILLE
CAMBRIDGE	83	81		98	187	124	80	155	128	145	185	63	49	142	128	228	50	23
CHILLICOTHE	184	57	98		108	199	47	77	69	40	89	35	107	44	85	189	147	75
CINCINNATI	243	152	187	108		259	110	53	154	61	148	134	236	104	79	209	236	164
COLUMBUS	129	74	80	47	110		144		70	114	69	135	30	129	91	44	148	130	58
DAYTON	198	146	155	77	53	213	70		145	57	160	102	204	116	26	156	204	132

DRIVING DISTANCES IN MILES

SEE ALSO MILEAGE AND DRIVING TIME MAP ON PAGE 144

Insets: YOUNGSTOWN–WARREN, COLUMBUS, SPRINGFIELD

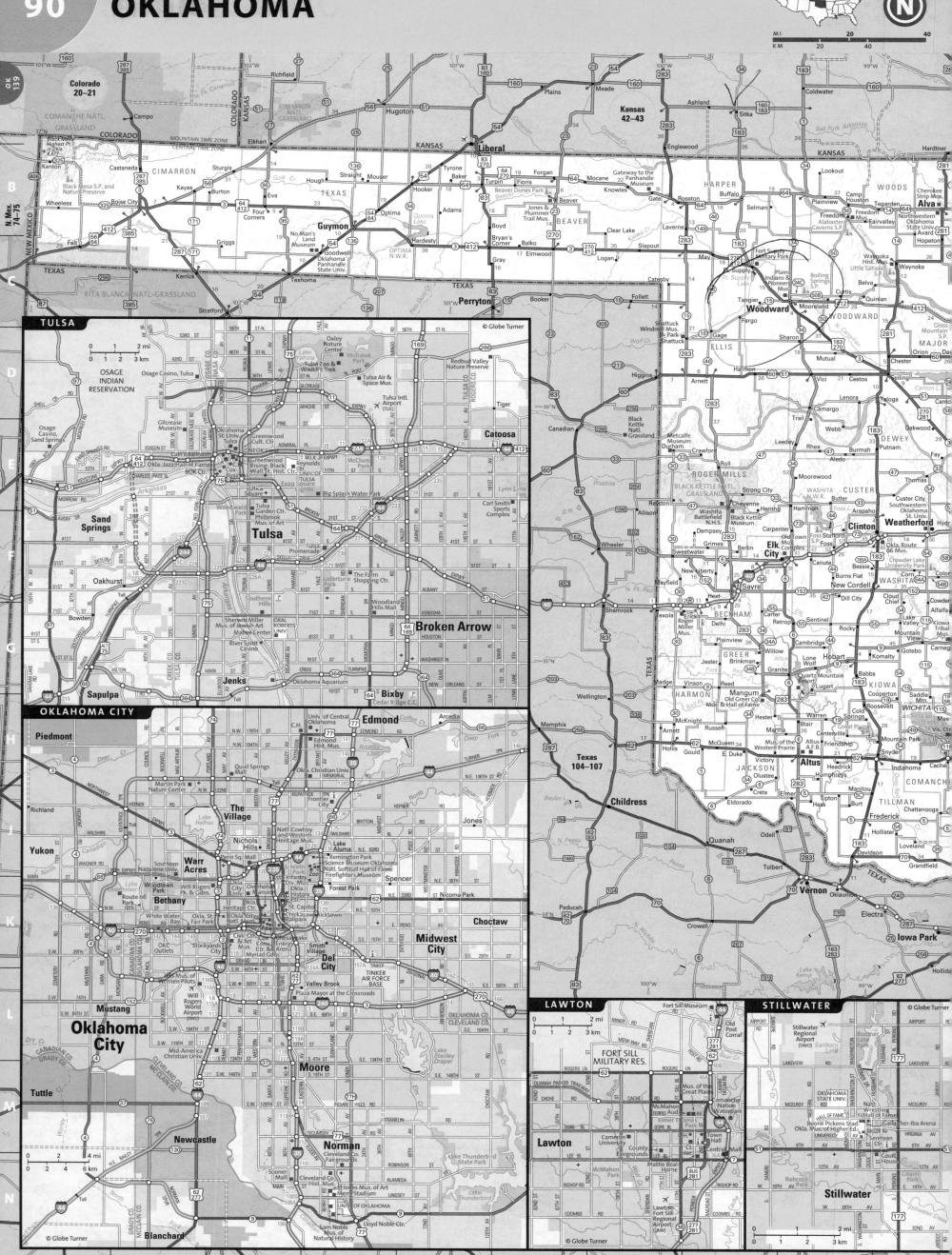

N

MI 20 40
KM 20 40

Colorado 20-21

Kansas 42-43

N. Mex. 74-75

Texas 104-107

TULSA

OSAGE INDIAN RESERVATION

Tulsa

Broken Arrow

Jenks

Sapulpa

Sand Springs

Oakhurst

Bowden

OKLAHOMA CITY

Piedmont

Edmond

Arcadia

The Village

Yukon

Warr Acres

Nichols Hills

Bethany

Oklahoma City

Mustang

Tuttle

Del City

Midwest City

Choctaw

Moore

Newcastle

Norman

Blanchard

LAWTON

Fort Sill Military Res.

Lawton

STILLWATER

Stillwater

Woodward

Elk City

Weatherford

Clinton

Sayre

New Cordell

Altus

Childress

Vernon

Iowa Park

Electra

Quanah

Frederick

© Globe Turner

	ARDMORE	BARTLESVILLE	DALLAS TX.	DURANT	ELK CITY	ENID	FORT SMITH AR	GUYMON	HUGO	LAWTON	MC ALESTER	MIAMI	MUSKOGEE	OKLAHOMA CITY	PONCA CITY	STILLWATER	TULSA	WOODWARD
ENID	183	141	292	238	148		242	219	282	142	210	207	168	84	69	66	117	88
LAWTON	103	243	197	158	115	142	270	297	254		211	283	223	85	192	152	194	175
MC ALESTER	117	141	169	77	245	210	114	407	75	211		160	68	133	186	154	93	276
OKLAHOMA CITY	99	157	209	154	112	84	191	274	205	85	133	198	144		107	67	109	143
TULSA	206	48	259	168	221	117	125	336	165	194	93	91	52	109	93	71		205

DRIVING DISTANCES IN MILES   SEE ALSO MILEAGE AND DRIVING TIME MAP ON PAGE 144

N

MI 20 40
KM 20 40

Washington 114–115

California 12–15

PACIFIC OCEAN

Major cities and places shown on the map include:

Astoria, Warrenton, Seaside, Longview, Kelso, Toppenish, Grandview, Sunnyside, Scappoose, St. Helens, Vancouver, Portland, Camas, Washougal, Hood River, The Dalles, Goldendale, Forest Grove, Hillsboro, Beaverton, Tigard, Lake Oswego, Gresham, Sandy, Tillamook, Newberg, Oregon City, Canby, Wilsonville, Molalla, McMinnville, Woodburn, Mount Angel, Silverton, Lincoln City, Dallas, Keizer, Salem, Monmouth, Stayton, Madras, Newport, Corvallis, Albany, Lebanon, Sweet Home, Prineville, Redmond, Florence, Junction City, Eugene, Springfield, Bend, Cottage Grove, Reedsport, Oakridge, North Bend, Coos Bay, Bandon, Sutherlin, Roseburg, Winston, Myrtle Creek, Canyonville, La Pine, Crescent, Chemult, Gilchrist, Silver Lake, Port Orford, Gold Beach, Grants Pass, Central Point, Medford, White City, Talent, Ashland, Klamath Falls, Altamont, Lakeview, Brookings, Crescent City, Yreka

TILLAMOOK STATE FOREST, CLATSOP STATE FOREST, SIUSLAW NATIONAL FOREST, WILLAMETTE NATIONAL FOREST, MOUNT HOOD NATIONAL FOREST, DESCHUTES NATIONAL FOREST, UMPQUA NATIONAL FOREST, ROGUE RIVER NATIONAL FOREST, SISKIYOU NATIONAL FOREST, FREMONT-WINEMA NATIONAL FOREST, OCHOCO NATIONAL FOREST, CRATER LAKE NATIONAL PARK

CASCADE RANGE, COAST RANGES, KLAMATH MTS., CALAPOOYA MTS., OCHOCO MTS., PAULINA MOUNTAINS

YAKAMA IND. RES., WARM SPRINGS IND. RES., GRAND RONDE COMMUNITY

Counties: CLATSOP, COLUMBIA, TILLAMOOK, WASHINGTON, MULTNOMAH, CLACKAMAS, YAMHILL, POLK, MARION, LINCOLN, BENTON, LINN, LANE, DOUGLAS, COOS, CURRY, JOSEPHINE, JACKSON, KLAMATH, DESCHUTES, CROOK, JEFFERSON, WASCO, SHERMAN, GILLIAM, WHEELER

	ASTORIA	BAKER CITY	BEND	BURNS	COOS BAY	CORVALLIS	CRATER LAKE N.P.	EUGENE	GRANTS PASS	KLAMATH FALLS	LAKEVIEW	MEDFORD	NEWPORT	ONTARIO	PENDLETON	PORTLAND	SALEM	THE DALLES
BEND	252	228		142	227	128	110	115	196	139	177	178	183	272	246	158	134	137
EUGENE	216	423	115	257	105	46	146		137	175	265	164	101	491	328	112	65	198
MEDFORD	375	406	178	311	170	205	80	164	28	76	171		260	441	487	271	224	357
PENDLETON	307	96	246	195	440	298	355	328	460	385	335	487	328	164		212	265	131
PORTLAND	97	307	158	299	224	82	253	112	244	282	335	287	116	375	212		48	82

DRIVING DISTANCES IN MILES   SEE ALSO MILEAGE AND DRIVING TIME MAP ON PAGE 144

© Globe Turner

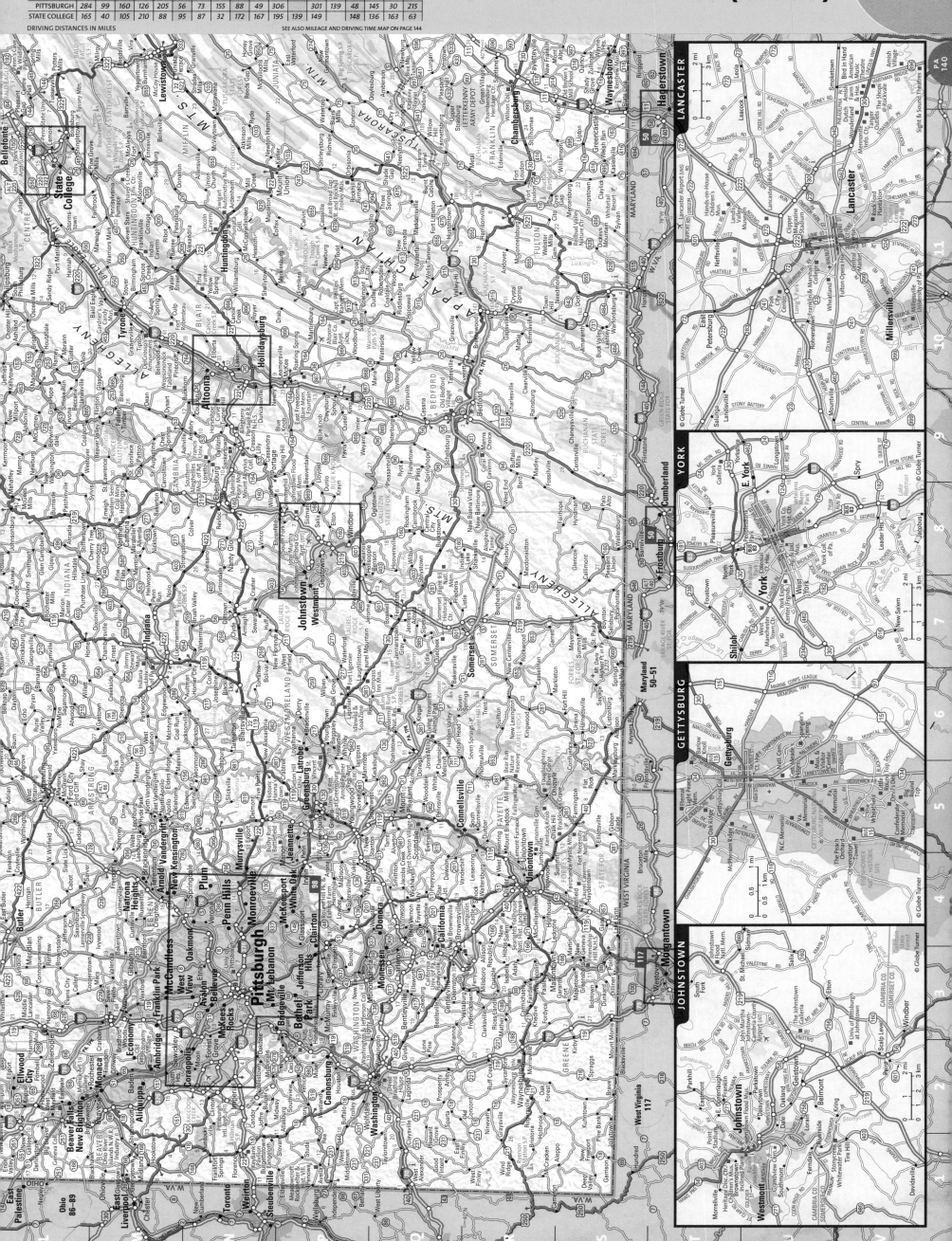

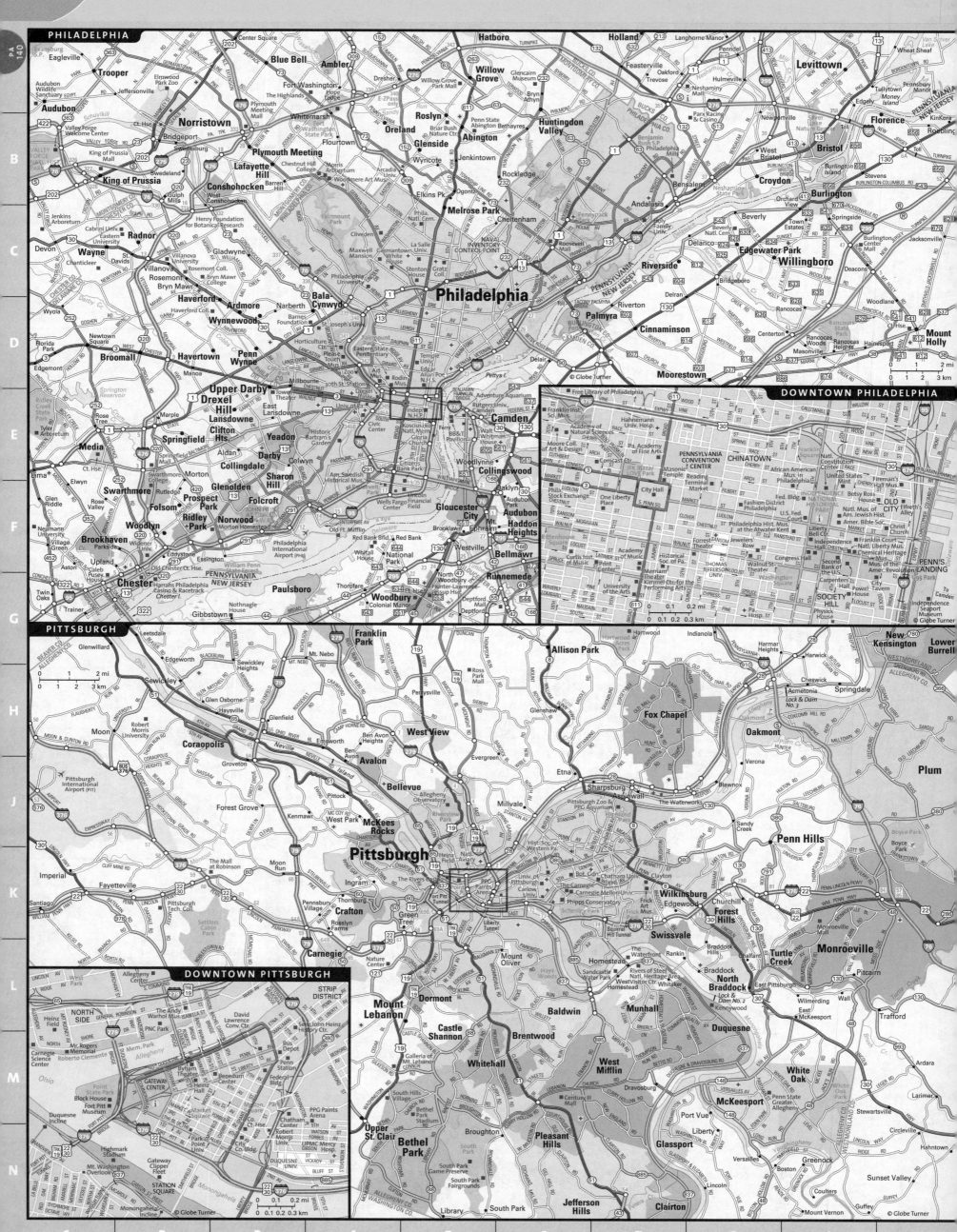

	BOSTON, MA	BRISTOL	EAST GREENWICH	FALL RIVER, MA	HOPE VALLEY	KINGSTON	NEWPORT	PROVIDENCE	WARWICK	WESTERLY	WICKFORD	WOONSOCKET
NEWPORT	73	14	20	20	28	17		33	27	41	13	46
PROVIDENCE	52	16	12	17	30	29	33		12	46	20	16
WARWICK	63	25	6	26	24	23	27	12		40	14	26
WESTERLY	97	60	41	61	17	26	41	46	40		34	59
WOONSOCKET	52	30	26	31	43	42	46	16	26	59	33	

DRIVING DISTANCES IN MILES   SEE ALSO MILEAGE AND DRIVING TIME MAP ON PAGE 144

PROVIDENCE

NEWPORT

Connecticut 22–23

Massachusetts 54–55

ATLANTIC OCEAN

© Globe Turner

	AUGUSTA, GA	CHARLOTTE, NC	CHARLESTON	COLUMBIA	FLORENCE	GREENVILLE	HILTON HEAD ISLAND	MYRTLE BEACH	ROCK HILL	SAVANNAH, GA	SPARTANBURG	SUMTER
CHARLESTON	142	204		110	127	205	95	92	183	107	200	100
COLUMBIA	70	91	110		80	97	152	146	70	159	92	45
FLORENCE	147	107	127	80		174	170	66	115	176	169	39
GREENVILLE	110	96	205	97	174		248	241	88	255	30	142
MYRTLE BEACH	213	173	92	146	92	241		190	181	235	93

DRIVING DISTANCES IN MILES

SEE ALSO MILEAGE AND DRIVING TIME MAP ON PAGE 144

MI 20 40
KM 20 40

© Globe Turner

SEE ALSO MILEAGE AND DRIVING TIME MAP ON PAGE 144

DRIVING DISTANCES IN MILES

	ABERDEEN	BELLE FOURCHE	BROOKINGS	HOT SPRINGS	HURON	MITCHELL	MOBRIDGE	PIERRE	RAPID CITY	SIOUX FALLS	WATERTOWN	YANKTON
ABERDEEN		310	150	412	90	146	99	160	357	204	98	231
PIERRE	160	247	188	247	115	155	107		193	226	189	240
RAPID CITY	357	56	390	46	313	275	243	193		346	436	360
SIOUX FALLS	204	401	57	401	127	73	303	226	346		103	80
WATERTOWN	98	360	49	490	86	162	196	189	436	103		179

	BRISTOL	CHATTANOOGA	CLARKSVILLE	COLUMBIA	COOKEVILLE	DYERSBURG	FAYETTEVILLE	GATLINBURG	JACKSON	JOHNSON CITY	KNOXVILLE	MANCHESTER	MEMPHIS	MORRISTOWN	MURFREESBORO	NASHVILLE	OAK RIDGE	UNION CITY
CHATTANOOGA	233		177	158	89	308	97	156	262	222	116	69	346	164	102	131	110	311
JOHNSON CITY	24	222	336	337	213	469	317	108	423		107	289	507	48	290	126	47	471
KNOXVILLE	117	116	230	231	107	363	211	40	317	107		183	401	48	179	184	24	365
MEMPHIS	518	346	213	210	296	81	268	441	91	507	401	270		449	246	215	383	113
NASHVILLE	301	131	46	49	79	178	91	223	132	290	184	64	215	232	31		166	181

DRIVING DISTANCES IN MILES

SEE ALSO MILEAGE AND DRIVING TIME MAP ON PAGE 144

KNOXVILLE

CHATTANOOGA

NORTHEASTERN TENNESSEE

© Globe Turner

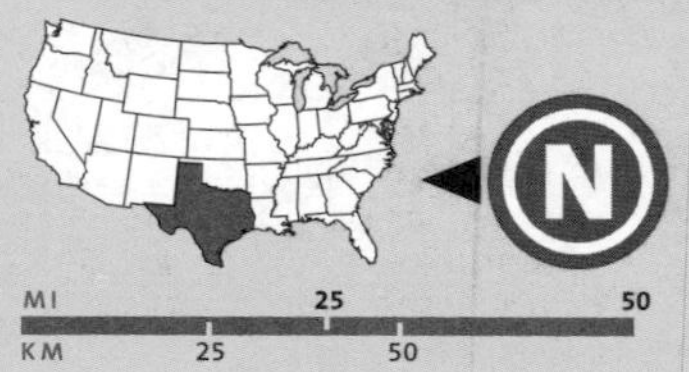

MI 25 50
KM 25 50

TX 140

Oklahoma 90–91

OKLAHOMA

NEW MEXICO

New Mexico 74–75

Amarillo
Lubbock
Abilene
Snyder
Sweetwater
Childress
Pampa
Borger
Perryton
Guymon
Dumas
Dalhart
Canyon
Hereford
Plainview
Tulia
Littlefield
Levelland
Brownfield
Seminole
Lamesa
Andrews
Slaton
Clovis
Portales
Tucumcari
Lovington
Hobbs
Midland
Odessa
West Odessa
El Paso
Ciudad Juárez
Las Cruces
Carlsbad
Canutillo
Sunland Park

PALO DURO CANYON

FORT BLISS MILITARY RESERVATION
BIGGS ARMY AIRFIELD
FRANKLIN MOUNTAINS
FRANKLIN MOUNTAINS STATE PARK

UNITED STATES
MEXICO
TEXAS
NEW MEXICO

© Globe Turner

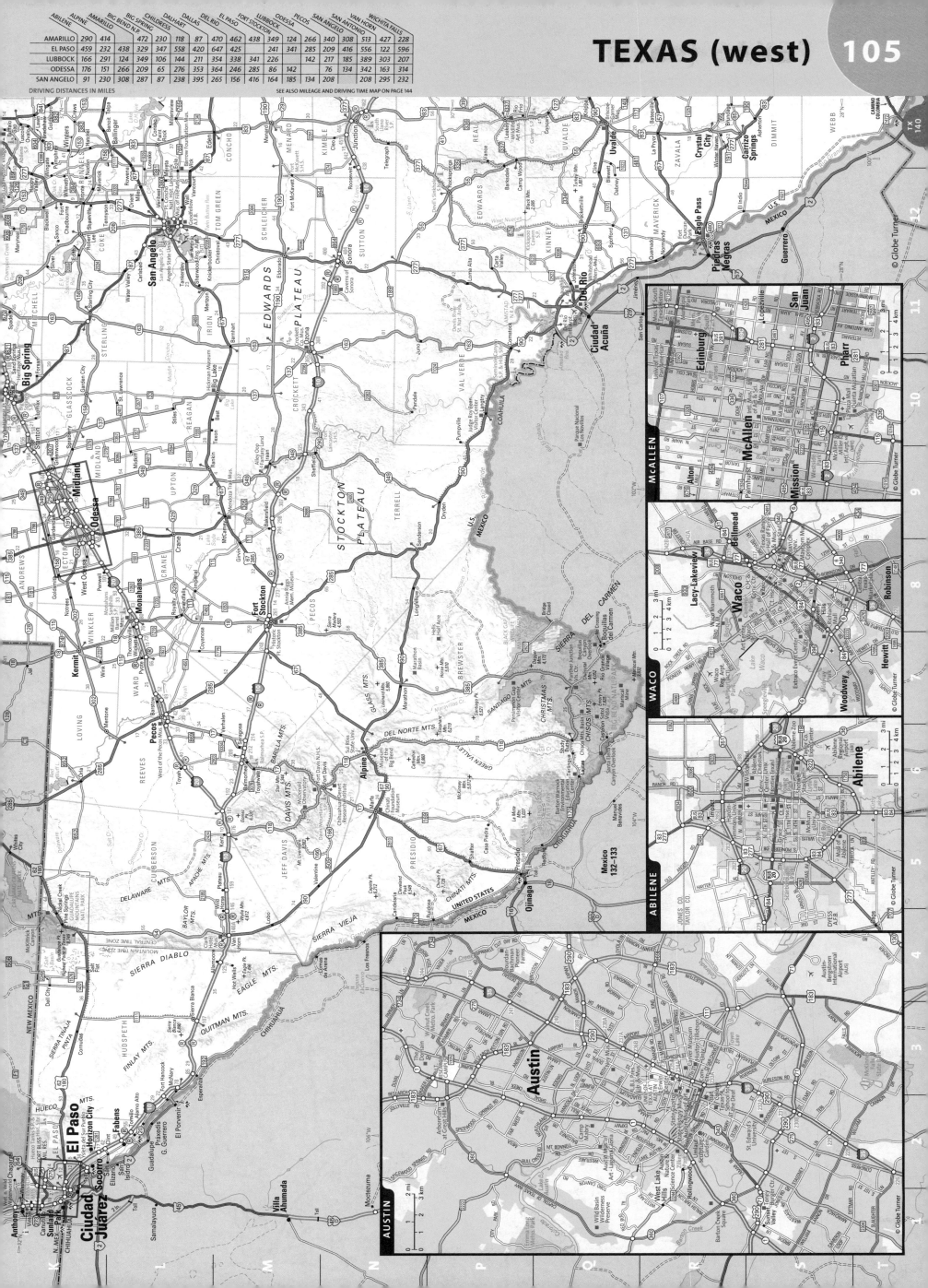

	ABILENE	ALPINE	AMARILLO	BIG BEND N.P.	BIG SPRING	CHILDRESS	DALHART	DALLAS	DEL RIO	EL PASO	FORT STOCKTON	LUBBOCK	ODESSA	PECOS	SAN ANGELO	SAN ANTONIO	VAN HORN	WICHITA FALLS
AMARILLO	290	414		472	230	118	87	470	462	438	349	124	266	340	308	513	427	228
EL PASO	459	232	438	329	347	558	420	647	425		241	341	285	209	416	556	122	596
LUBBOCK	166	291	124	349	106	144	211	354	338	341	226		142	217	185	389	303	207
ODESSA	176	151	266	209	65	276	353	364	246	285	86	142		76	134	342	163	314
SAN ANGELO	91	230	308	287	87	278	395	265	156	416	164	185	134	128		208	295	232

DRIVING DISTANCES IN MILES   SEE ALSO MILEAGE AND DRIVING TIME MAP ON PAGE 144

MI 25 50
KM 25 50

N

BEAUMONT

TX 140

© Globe Turner

Major cities and places: Texarkana, Shreveport, Bossier City, Marshall, Longview, Carthage, Nacogdoches, Lufkin, Diboll, Center, Jacksonville, Tyler, Henderson, Rusk, Crockett, Huntsville, College Station, Bryan, Palestine, Athens, Corsicana, Waco, Temple, Killeen, Copperas Cove, Harker Heights, Belton, Cameron, Marlin, Mexia, Paris, Sherman, Denison, Greenville, Commerce, Sulphur Springs, Mt. Pleasant, Atlanta, Dallas, Fort Worth, Arlington, Irving, Garland, Mesquite, Plano, McKinney, Allen, Frisco, Denton, Decatur, Bowie, Gainesville, Wichita Falls, Iowa Park, Burkburnett, Vernon, Childress, Lawton, Frederick, Altus, Duncan, Ardmore, Durant, Hugo, Idabel, Broken Bow, Antlers, Austin, Round Rock, Cedar Park, Georgetown, Taylor, Elgin, Bastrop, Giddings, Brenham, Navasota, Conroe, The Woodlands, Cleveland, Livingston, Jasper, Kirbyville, Silsbee, Beaumont, Orange, Vidor, Lumberton, DeRidder, Leesville, Many, Mansfield, Minden, Springhill, Magnolia, Fredericksburg, Brady, Brownwood, Coleman, Abilene, Breckenridge, Mineral Wells, Stephenville, Cleburne, Waxahachie, Ennis, Hillsboro, Gatesville, Lampasas, Burnet

Oklahoma 90-91
Arkansas 10-11
Louisiana 46-47

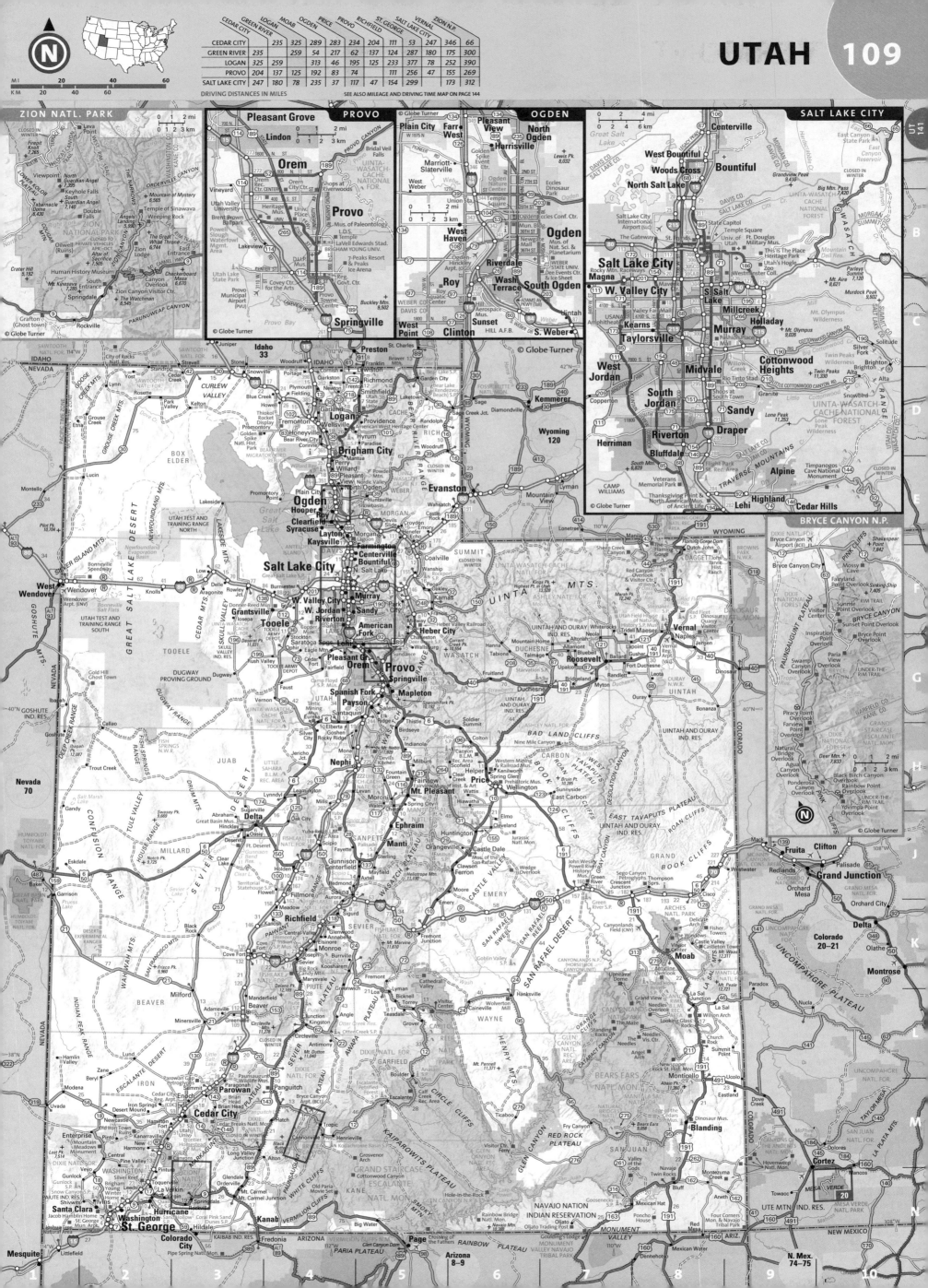

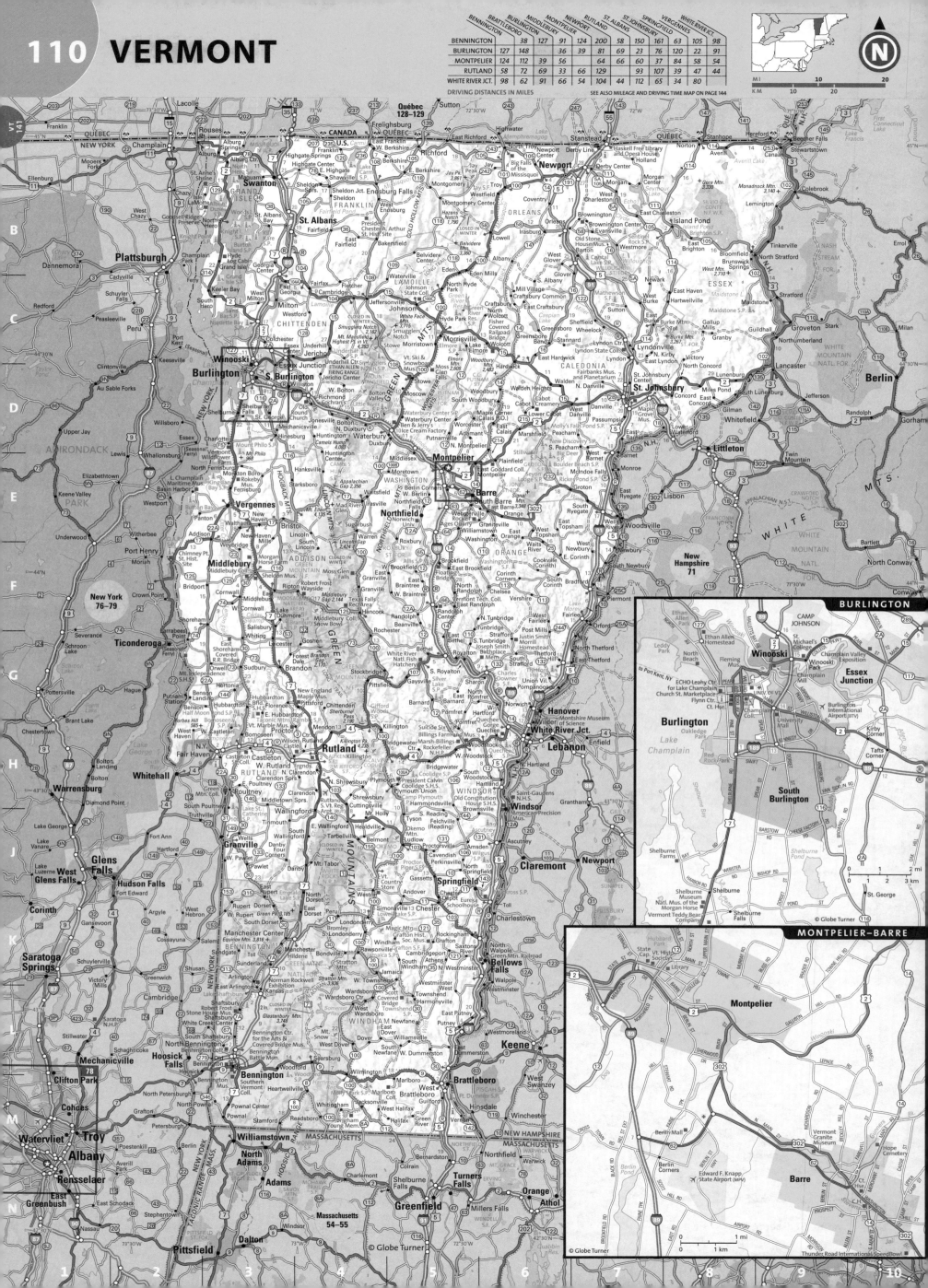

	BENNINGTON	BRATTLEBORO	BURLINGTON	MIDDLEBURY	MONTPELIER	NEWPORT	RUTLAND	ST. ALBANS	ST. JOHNSBURY	SPRINGFIELD	VERGENNES	WHITE RIVER JCT.
BENNINGTON		38	127	91	124	200	58	150	161	63	105	98
BURLINGTON	127	148		36	39	81	69	23	76	120	22	91
MONTPELIER	124	112	39	56		64	66	60	37	84	58	54
RUTLAND	58	72	69	33	66	129		93	107	39	47	44
WHITE RIVER JCT.	98	62	91	66	54	104	44	112	65	34	80	

DRIVING DISTANCES IN MILES

SEE ALSO MILEAGE AND DRIVING TIME MAP ON PAGE 144

BURLINGTON

MONTPELIER–BARRE

© Globe Turner

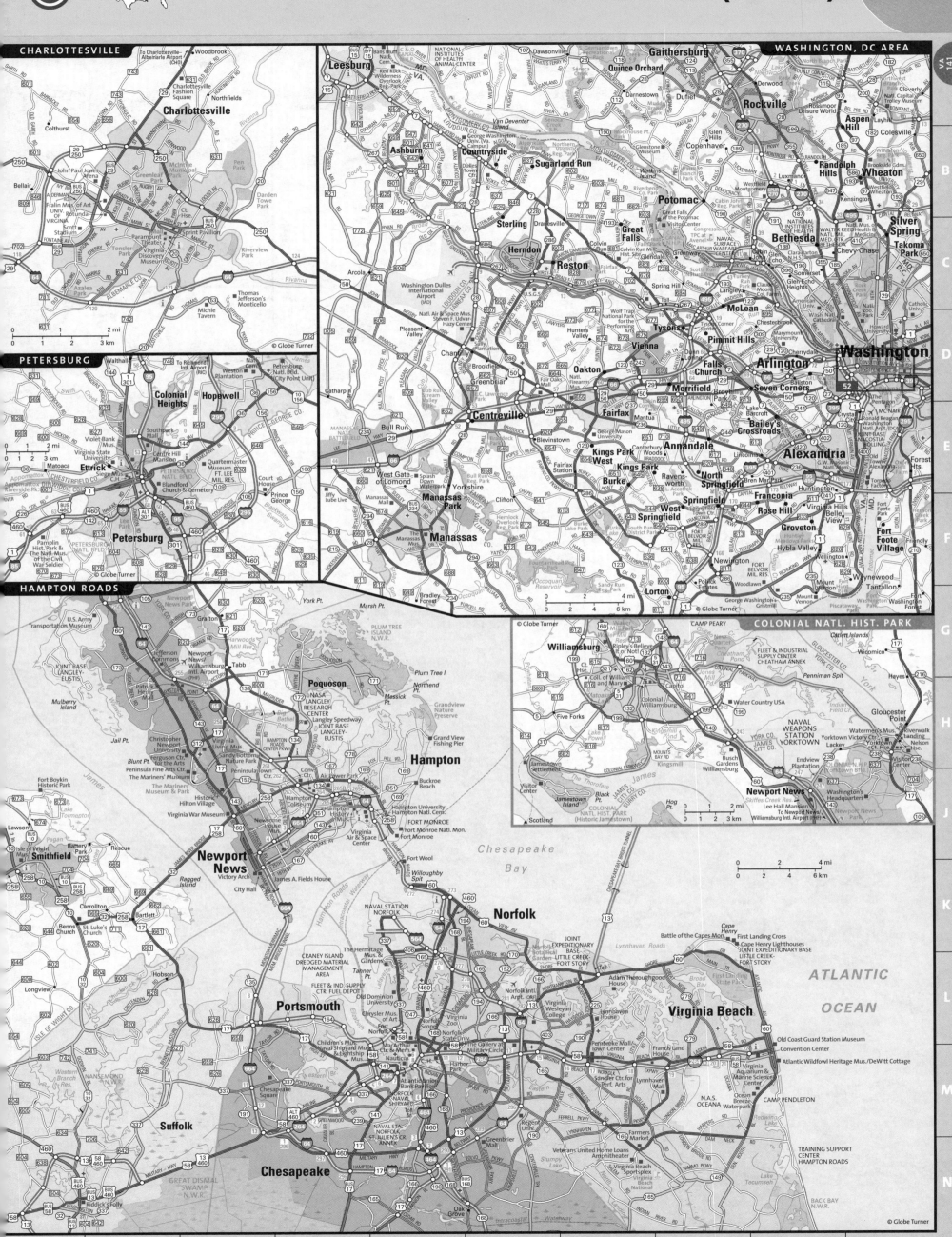

CHARLOTTESVILLE

WASHINGTON, DC AREA

PETERSBURG

HAMPTON ROADS

COLONIAL NATL. HIST. PARK

© Globe Turner

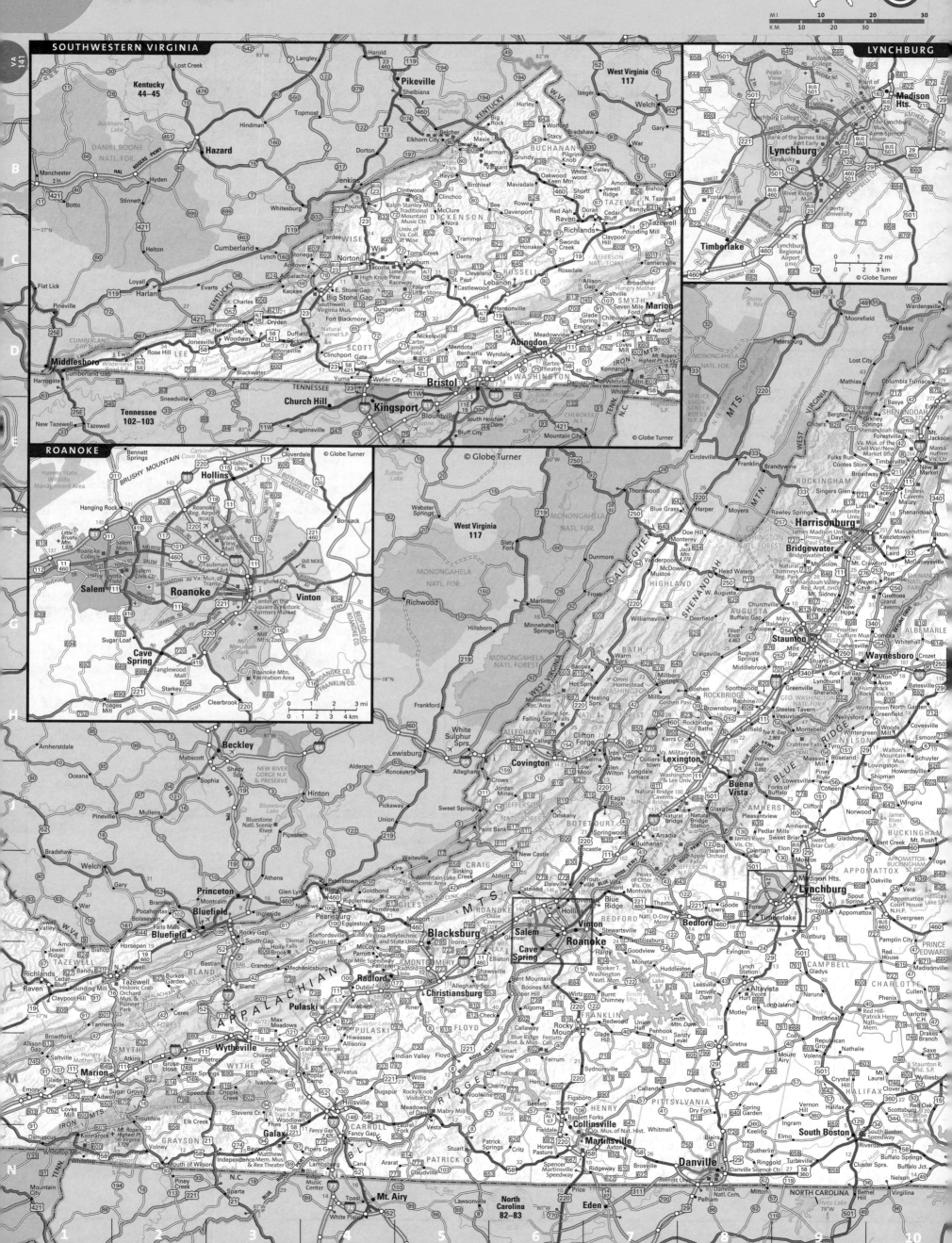

	BRISTOL	CHARLOTTESVILLE	CHINCOTEAGUE	DANVILLE	EMPORIA	FREDERICKSBURG	HARRISONBURG	LEXINGTON	LYNCHBURG	NORFOLK	PETERSBURG	RICHMOND	ROANOKE	VIRGINIA BEACH	WASHINGTON DC	WILLIAMSBURG	WINCHESTER	WYTHEVILLE
CHARLOTTESVILLE	258		255	131	140	70	62	69	67	164	95	71	123	180	118	122	100	187
NORFOLK	420	164	101	191	78	145	225	231	203		79	91	285	18	196	43	235	350
RICHMOND	327	71	182	160	69	57	131	138	114	91	24		192	107	108	49	147	256
ROANOKE	149	123	377	83	180	192	117	54	55	285	216	192		301	245	243	184	78
WASHINGTON, DC	380	118	158	268	177	54	131	191	183	196	133	108	245	212		153	92	309

DRIVING DISTANCES IN MILES

SEE ALSO MILEAGE AND DRIVING TIME MAP ON PAGE 144

RICHMOND

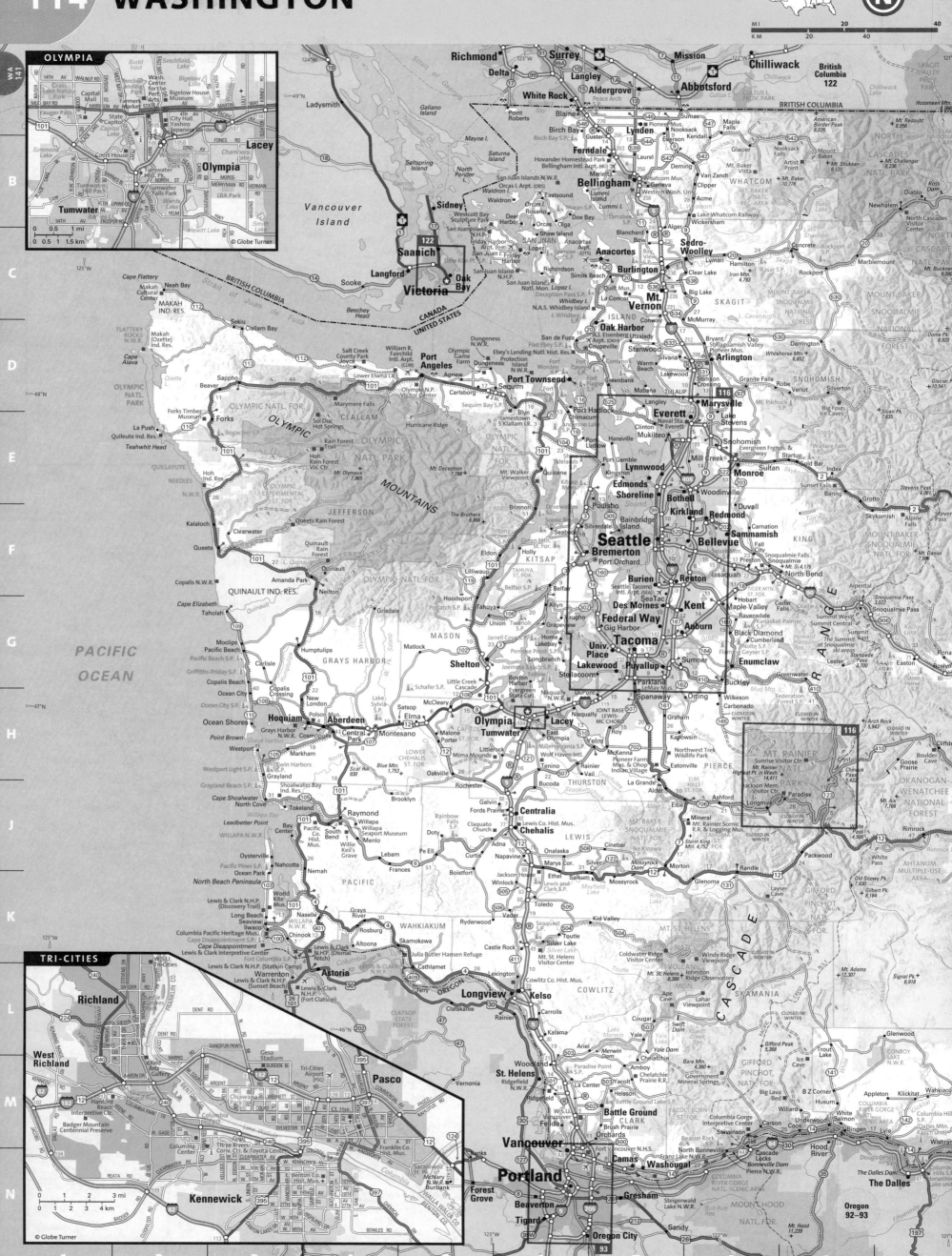

OLYMPIA

Budd Inlet
Setchfield Ave
WALNUT RD
14TH AV
Wash. Center for the Performing Arts
Capital Mall
State Capitol
Capitol Lake
Tumwater Falls Park
Lacey
Olympia
Tumwater
© Globe Turner

TRI-CITIES

Richland
West Richland
Kennewick
Pasco
Columbia Center
Tri-Cities Airport (PSC)
© Globe Turner

PACIFIC OCEAN

British Columbia 122

British Columbia

Richmond
Surrey
Delta
Langley
White Rock
Aldergrove
Abbotsford
Chilliwack
Mission
Blaine
Lynden
Ferndale
Bellingham
Sedro-Woolley
Anacortes
Burlington
Mt. Vernon
Oak Harbor
Saanich
Victoria
Oak Bay
Port Angeles
Port Townsend
Sequim
Marysville
Everett
Arlington
Forks
Mukilteo
Lynnwood
Edmonds
Shoreline
Monroe
Bothell
Kirkland
Redmond
Woodinville
Duvall
Seattle
Bremerton
Bellevue
Sammamish
Port Orchard
Renton
North Bend
Burien
Issaquah
SeaTac
Kent
Des Moines
Auburn
Federal Way
Enumclaw
Tacoma
Univ. Place
Puyallup
Lakewood
Sumner
Buckley
Shelton
Parkland
Spanaway
Orting
Hoquiam
Aberdeen
Central Park
Montesano
Elma
McCleary
Olympia
Tumwater
Lacey
Yelm
Ocean Shores
Ocean City
Copalis Beach
Mt. Rainier Natl. Park
Paradise
Centralia
Chehalis
Raymond
South Bend
Astoria
Warrenton
Longview
Kelso
St. Helens
Woodland
Battle Ground
Vancouver
Camas
Washougal
Portland
Beaverton
Gresham
Tigard
Oregon City
The Dalles
Hood River

PACIFIC OCEAN

Oregon 92-93

	Aberdeen	Bellingham	Everett	Kennewick	Lewiston, ID	Longview	Mount Rainier N.P.	Okanogan	Olympia	Port Angeles	Portland, OR	Pullman	Seattle	Spokane	Tacoma	Walla Walla	Wenatchee	Yakima
BELLINGHAM	196		61	307	420	215	186	195	147	127*	261	390	88	360	122	353	185	221
OLYMPIA	49	147	86	275	387	68	73	282		117	114	358	56	327	27	320	196	188
SEATTLE	105	88	28	276	338	124	96	223	56	83*	170	309		278	31	271	148	140
SPOKANE	376	360	299	139	103	395	290	148	327	362*	351	73	278		303	167	171	203
YAKIMA	237	221	161	86	214	170	87	194	188	223*	187	233	140	203	164	132	115	

DRIVING DISTANCES IN MILES   * DISTANCE INCLUDES FERRY TRAVEL

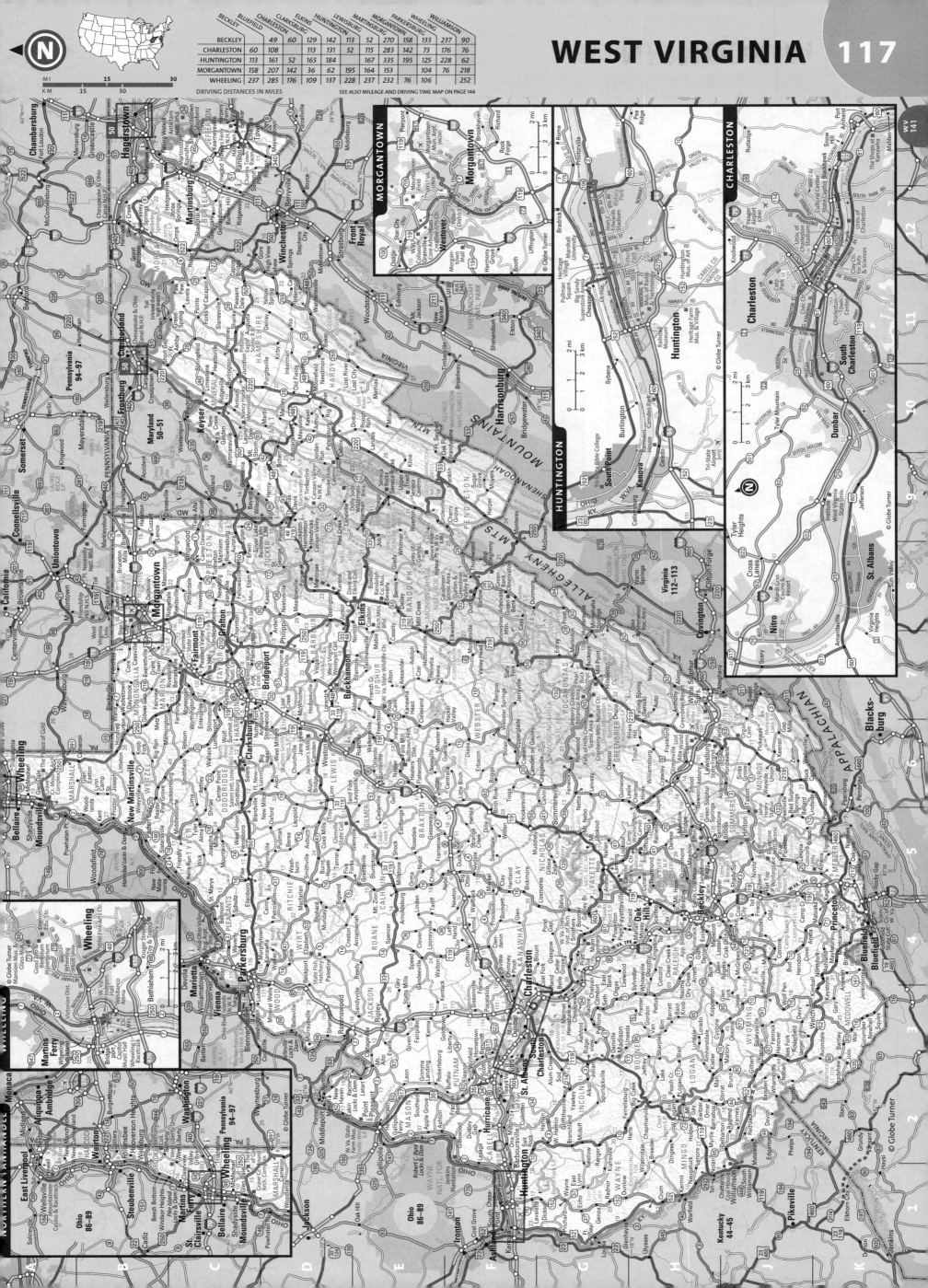

	BECKLEY	BLUEFIELD	CHARLESTON	CLARKSBURG	ELKINS	HUNTINGTON	LEWISBURG	MARTINSBURG	MORGANTOWN	PARKERSBURG	WHEELING	WILLIAMSON
BECKLEY		49	60	129	142	113	52	270	158	133	237	90
CHARLESTON	60	108		113	131	52	115	283	142	73	176	76
HUNTINGTON	113	161	52	165	184		167	335	195	125	228	62
MORGANTOWN	158	207	142	36	62	195	153	153		104	76	218
WHEELING	237	285	176	109	137	228	237	232	76	106		252

DRIVING DISTANCES IN MILES

SEE ALSO MILEAGE AND DRIVING TIME MAP ON PAGE 144

N

MI 20 40
KM 20 40

RACINE–KENOSHA

Caledonia
Mount Pleasant
Racine
Sturtevant
Somers
Kenosha
Pleasant Prairie

© Globe Turner

GREEN BAY

Howard
Hobart
Green Bay
Bellevue
Allouez
Ashwaubenon
De Pere

© Globe Turner

OSHKOSH

Winnebago
Oshkosh

© Globe Turner

APPLETON

Little Chute
Kaukauna
Kimberly
Combined Locks
Appleton
Menasha
Neenah
Fox Crossing

© Globe Turner

LAKE MICHIGAN

LAKE SUPERIOR

APOSTLE ISLANDS

MICH.

MINNESOTA

Michigan 56–57

Minnesota 60–61

Marinette
Menominee
Iron Mountain
Kingsford
Florence
Escanaba
Sturgeon Bay
Green Bay
Rhinelander
Antigo
Shawano
Merrill
Wausau
Weston
Rothschild
Stevens Point
Marshfield
Ashland
Park Falls
Rice Lake
Chippewa Falls
Altoona
Eau Claire
Menomonie
New Richmond
River Falls
Duluth
Superior
Spooner
St. Paul
Hastings
Red Wing

	APPLETON	ASHLAND	BELOIT	EAU CLAIRE	GREEN BAY	KENOSHA	LA CROSSE	MADISON	MARINETTE	MILWAUKEE	OSHKOSH	PORTAGE	PRAIRIE DU CHIEN	RHINELANDER	ST. PAUL MN	SHEBOYGAN	SUPERIOR	WAUSAU
EAU CLAIRE	178	163	225		192	284	81	176	211	246	180	147	151	151	83	227	155	99
GREEN BAY	31	245	189	192		154	205	135	54	115	50	120	234	124	268	61	308	93
MADISON	104	305	56	176	135	116	141		184	78	86	36	102	197	258	132	327	141
MILWAUKEE	105	375	74	246	115	39	211	78	170		87	106	180	239	328	54	397	211
WAUSAU	90	165	190	99	93	249	146	141	112	211	111	105	193	58	174	153	223	

DRIVING DISTANCES IN MILES

SEE ALSO MILEAGE AND DRIVING TIME MAP ON PAGE 144

	CASPER	CHEYENNE	CODY	EVANSTON	GILLETTE	JACKSON	LANDER	LARAMIE	RAWLINS	ROCK SPRINGS	SHERIDAN	YELLOWSTONE N.P.
CASPER		175	215	308	127	282	148	117	214	149	298	
CHEYENNE	175		390	353	243	433	276	52	151	260	324	455
JACKSON	282	433		181	195	412	163	383	283	177	376	80
ROCK SPRINGS	214	260	281	97	341	177	118	210	110		363	259
SHERIDAN	149	324	150	457	102	376	238	297	266	363		337

DRIVING DISTANCES IN MILES

SEE ALSO MILEAGE AND DRIVING TIME MAP ON PAGE 144

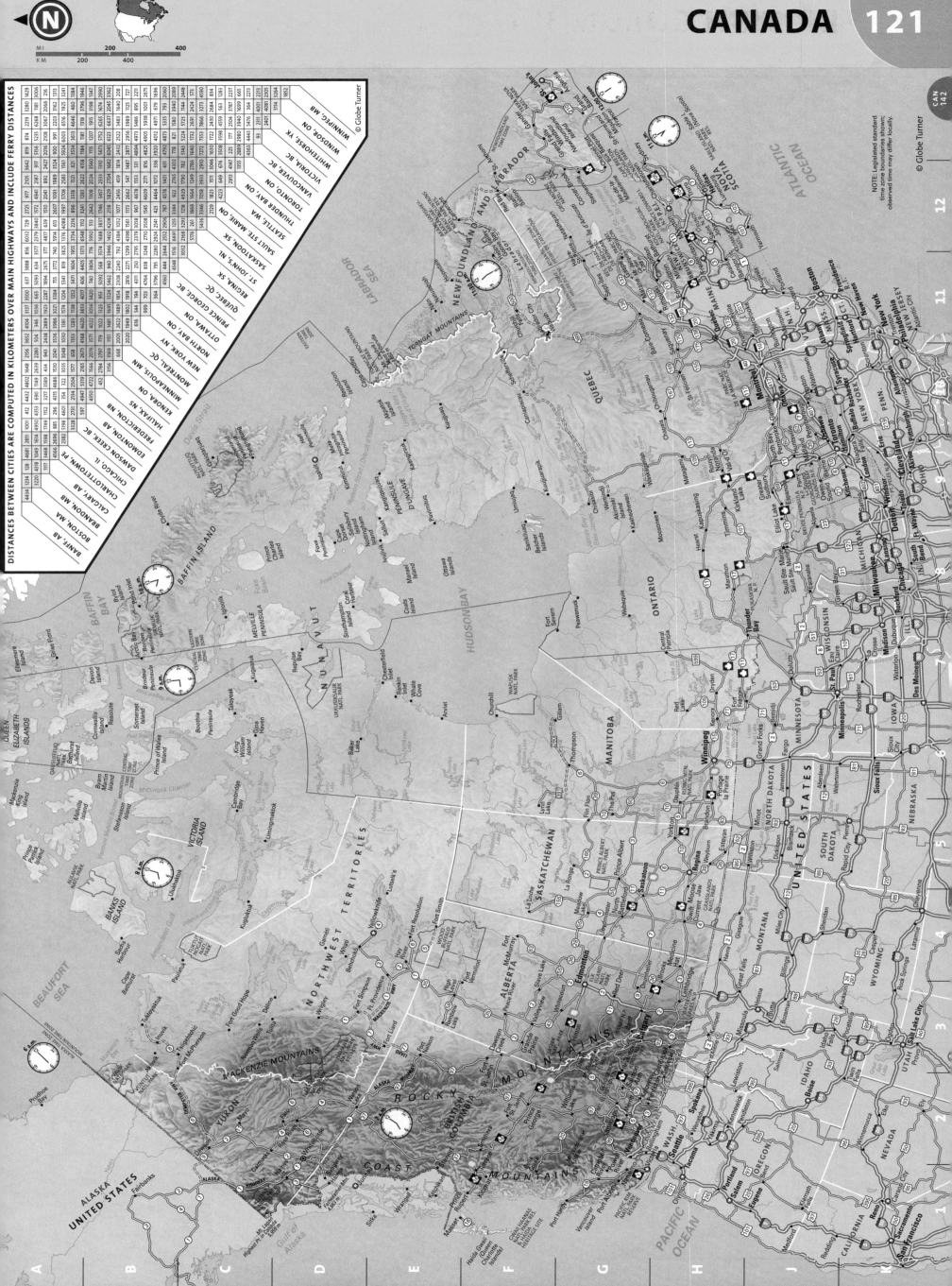

NOTE: Legislated standard time zone boundaries shown; observed time may differ locally.

CAN 142

	BANFF, AB CRANBROOK		DAWSON CREEK	JASPER, AB	KAMLOOPS	KELOWNA	NANAIMO	PRINCE GEORGE	PRINCE RUPERT	REVELSTOKE	VANCOUVER	VICTORIA
CRANBROOK	265		985	504	600	513	845*	838	1562	196	765	882*
KAMLOOPS	479	600	931	444		163	359*	525	1249	206	340	396*
KELOWNA	476	513	1094	597	163		383*	688	1409	192	378	420*
PRINCE GEORGE	637	838	406	376	525	688	784*		724	693	778	821*
VANCOUVER	819	765	1184	784	340	378	109*	778	1502	546		93*

DRIVING DISTANCES IN KILOMETERS   * DISTANCE INCLUDES FERRY TRAVEL

MI   40   80
KM   40   80

VANCOUVER

KAMLOOPS

VICTORIA

© Globe Turner

DISTANCES IN CANADA SHOWN IN KILOMETERS

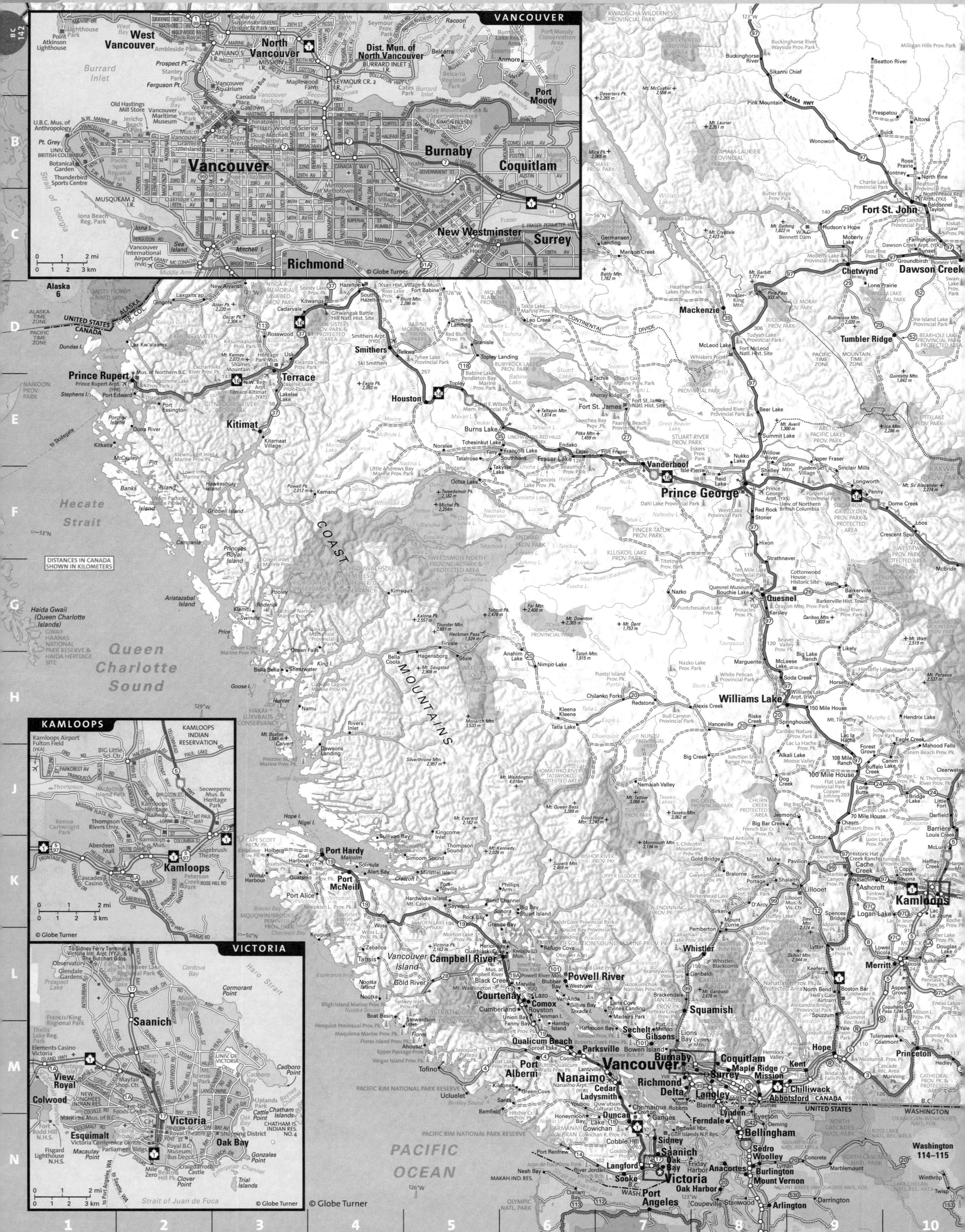

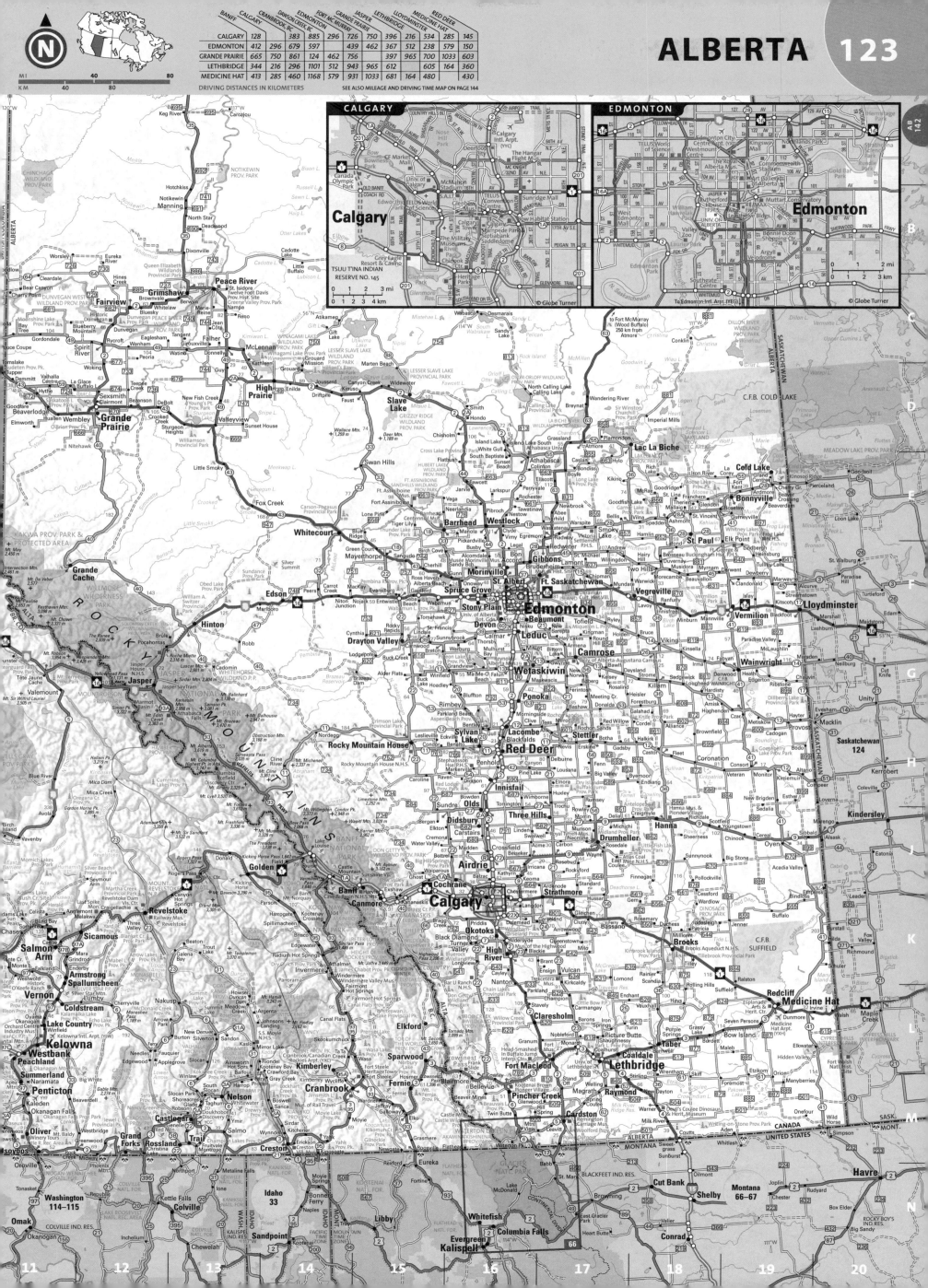

	BANFF	CALGARY	CRANBROOK BC	DAWSON CREEK BC	EDMONTON	FORT MC MURRAY	GRANDE PRAIRIE	JASPER	LETHBRIDGE	LLOYDMINSTER	MEDICINE HAT	RED DEER
CALGARY	128		383	885	296	750	726	396	216	534	285	145
EDMONTON	412	296	679	597		439	462	367	512	238	579	150
GRANDE PRAIRIE	665	750	861	124	462	756		397	965	700	1033	603
LETHBRIDGE	344	216	296	1101	512	943	965	612		605	164	360
MEDICINE HAT	413	285	460	1108	579	931	1033	681	164	480		430

DRIVING DISTANCES IN KILOMETERS

SEE ALSO MILEAGE AND DRIVING TIME MAP ON PAGE 144

	BRANDON, MB	ESTEVAN	FLIN FLON, MB	LLOYDMINSTER	MEDICINE HAT, AB	MOOSE JAW	NORTH BATTLEFORD	PRINCE ALBERT	REGINA	SASKATOON	SWIFT CURRENT	YORKTON
PRINCE ALBERT	670	570	375	336	618	356	196		374	141	408	391
REGINA	377	196	748	537	455	68	397	374		261	241	195
SASKATOON	639	455	508	275	486	224	137	141	261		267	331
SWIFT CURRENT	598	401	989	441	218	174	301	408	241	267		436
YORKTON	270	289	553	608	650	262	468	391	195	331	436	

DRIVING DISTANCES IN KILOMETERS   SEE ALSO MILEAGE AND DRIVING TIME MAP ON PAGE 144

MI [scale] 30   60
KM [scale] 30   60

SASKATOON

REGINA

© Globe Turner

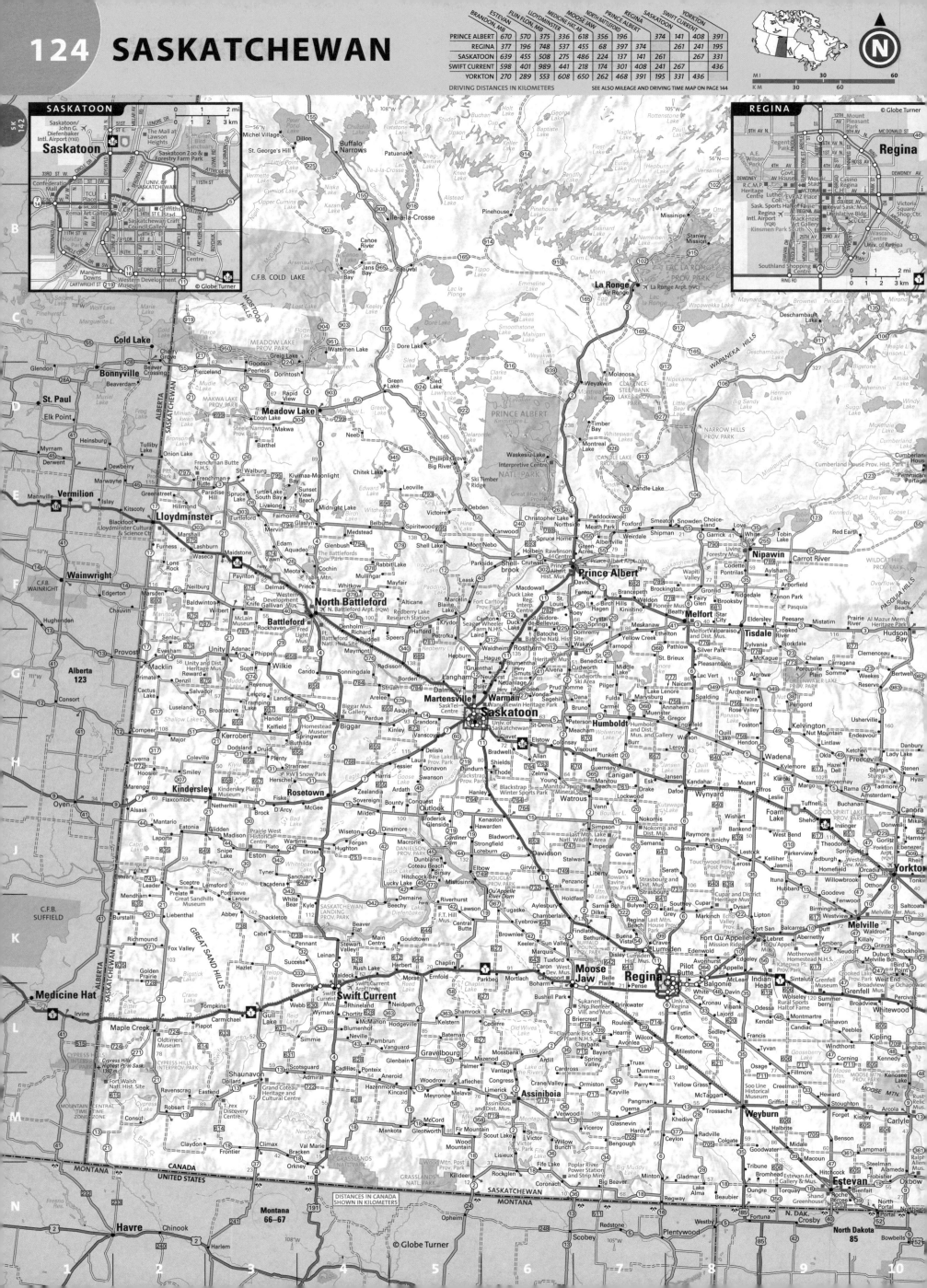

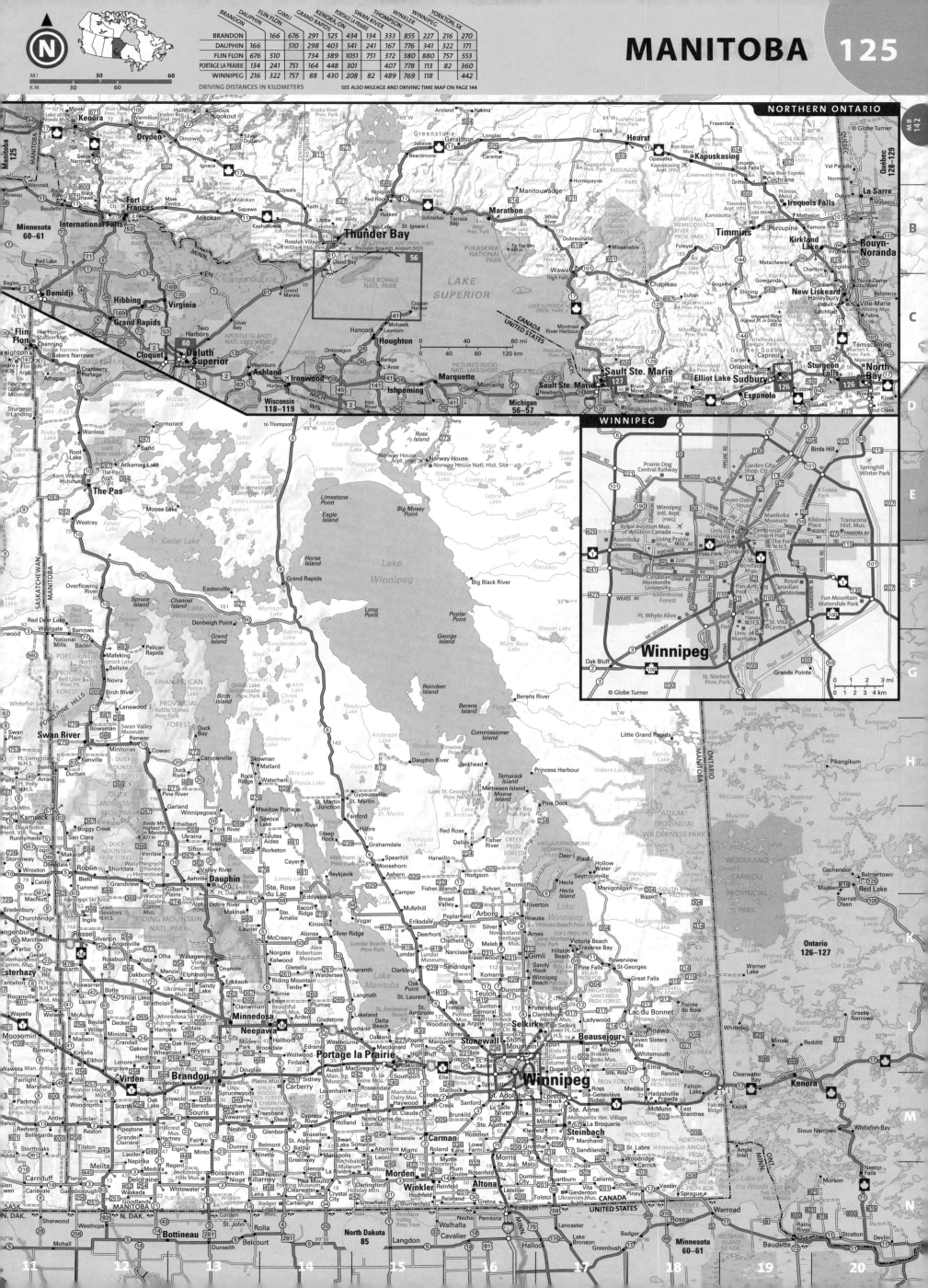

	BRANDON	DAUPHIN	FLIN FLON	GIMLI	GRAND RAPIDS	KENORA, ON	PORTAGE LA PRAIRIE	SWAN RIVER	THOMPSON	WINKLER	WINNIPEG	YORKTON, SK
BRANDON		166	676	291	525	434	134	333	855	227	216	270
DAUPHIN	166		510	298	403	541	241	167	776	341	322	171
FLIN FLON	676	510		734	389	1051	751	372	380	880	757	553
PORTAGE LA PRAIRIE	134	241	751	164	448	301		407	778	113	82	360
WINNIPEG	216	322	757	88	430	208	82	489	769	118		442

DRIVING DISTANCES IN KILOMETERS
SEE ALSO MILEAGE AND DRIVING TIME MAP ON PAGE 144

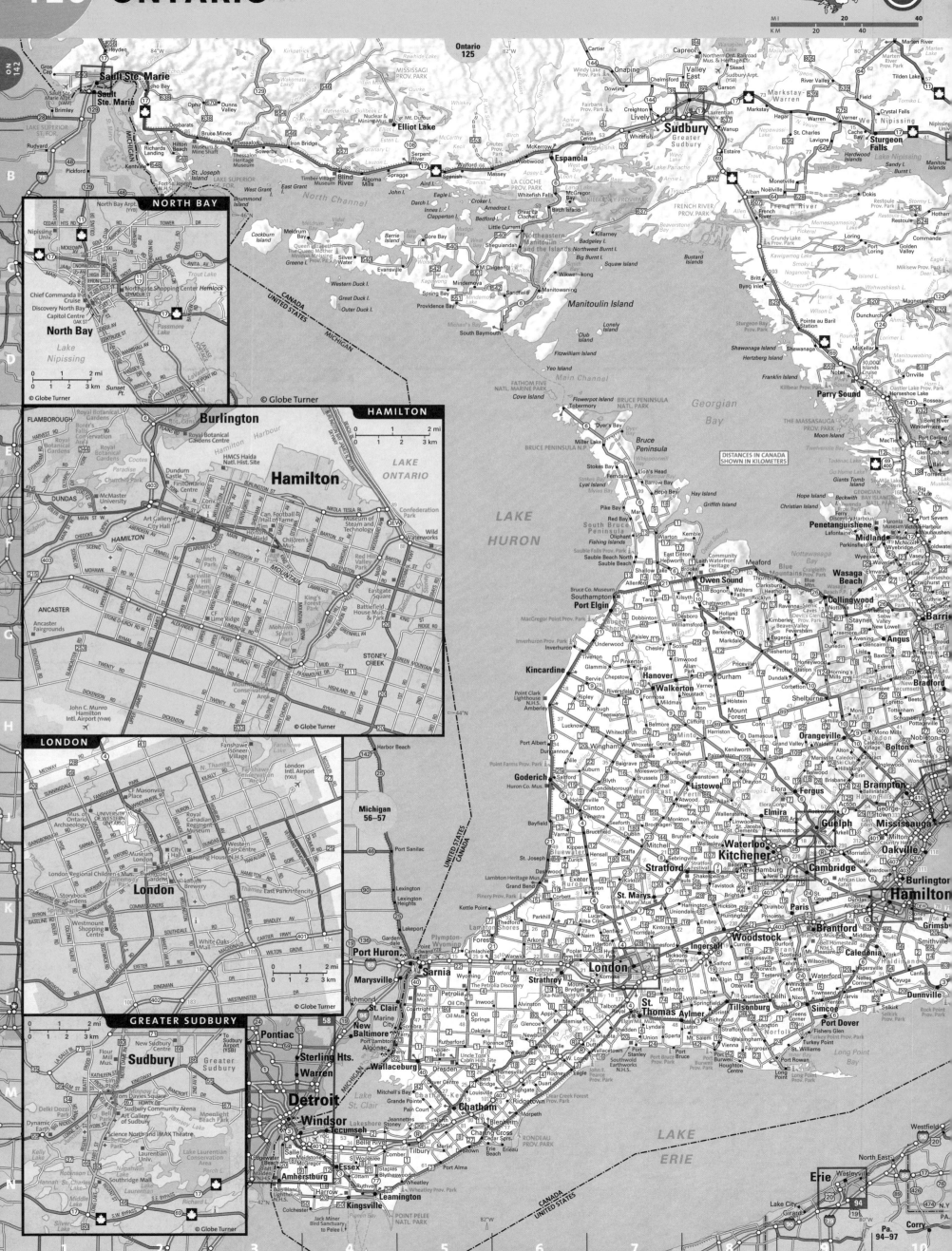

MI | 20 | 40
KM | 20 | 40

NORTH BAY

North Bay

Lake Nipissing

© Globe Turner

BURLINGTON

HAMILTON

Hamilton

LONDON

London

© Globe Turner

GREATER SUDBURY

Sudbury

Greater Sudbury

© Globe Turner

Ontario 125

Sault Ste. Marie

Elliot Lake

Espanola

Sudbury

Greater Sudbury

Sturgeon Falls

North Channel

Manitoulin Island

Parry Sound

LAKE HURON

Georgian Bay

Owen Sound

Port Elgin

Kincardine

Walkerton

Hanover

Goderich

Listowel

Fergus

Guelph

Brampton

Mississauga

Oakville

Waterloo

Kitchener

Cambridge

Hamilton

Stratford

Brantford

Woodstock

Port Huron

Sarnia

London

St. Thomas

Aylmer

Tillsonburg

Simcoe

Port Dover

Detroit

Windsor

Chatham

Leamington

LAKE ERIE

Erie

CANADA
UNITED STATES

DRIVING DISTANCES IN KILOMETERS

	BARRIE	HAMILTON	KENORA	KINGSTON	KITCHENER	LONDON	NIAGARA FALLS	NORTH BAY	OTTAWA	OWEN SOUND	PETERBOROUGH	SARNIA	SAULT STE. MARIE	SUDBURY	THUNDER BAY	TIMMINS	TORONTO	WINDSOR
LONDON	248	134	1926	434	105		227	499	613	208	309	169	818	995	1467	840	183	195
OTTAWA	442	504	1854	179	496	613	574	364		558	205	714	787	488	1401	705	431	793
SUDBURY	319	462	1407	600	453	570	533	124	488	435	404	673	299		948	290	407	751
THUNDER BAY	1219	1410	459	1548	1401	1467	1481	1072	1401	1335	1371	1621	649	948		1355	1355	1699
TORONTO	105	74	1814	251	104	183	145	336	431	183	127	285	674	407	1355	677		364

SEE ALSO MILEAGE AND DRIVING TIME MAP ON PAGE 144

GUELPH

KITCHENER/WATERLOO/CAMBRIDGE

DOWNTOWN TORONTO

SAULT STE. MARIE

TORONTO

Québec 128–129

New York 76–79

© Globe Turner

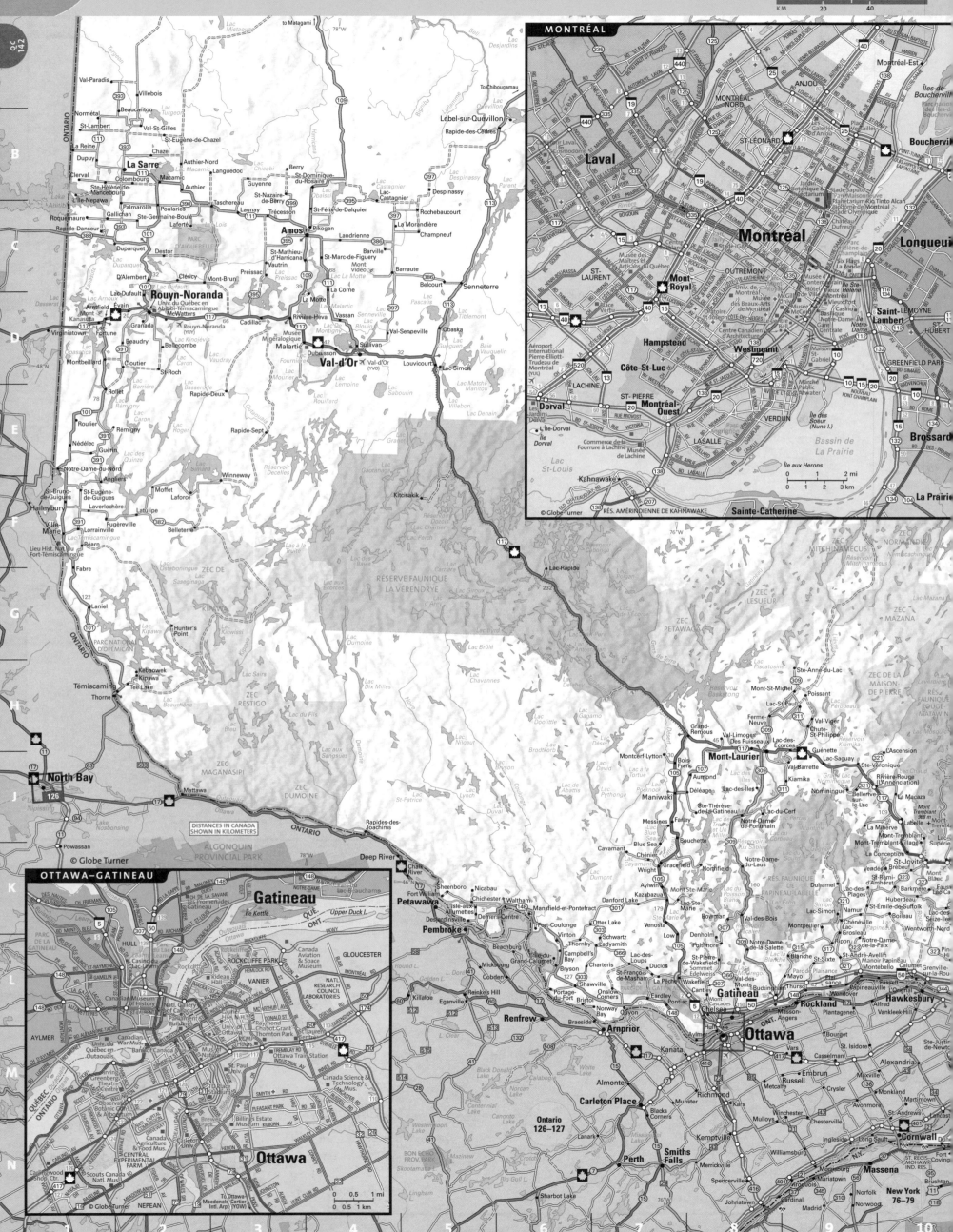

	BAIE-COMEAU	CHICOUTIMI	DRUMMONDVILLE	GASPÉ	MONT-LAURIER	MONTRÉAL	OTTAWA, ON	QUÉBEC	RIMOUSKI	RIVIÈRE-DU-LOUP	ROBERVAL	ROUYN-NORANDA	ST-GEORGES	SEPT-ÎLES	SHERBROOKE	SOREL	TROIS-RIVIÈRES	VICTORIAVILLE
MONTRÉAL	663	461	116	898	230		194	250	535	426	448	616	325	887	143	87	146	164
QUÉBEC	400	211	151	668	439	250	444		305	196	253	879	102	624	233	204	135	114
RIVIÈRE-DU-LOUP	230*	154*	328	472	656	426	620	196	109		249*	1042	272	454*	401	381	333	291
SHERBROOKE	633	444	82	873	373	143	337	233	510	401	417	759	148	857		142	158	97
TROIS-RIVIÈRES	535	346	68	804	376	146	340	135	441	333	296	762	214	759	158	82		65

DRIVING DISTANCES IN KILOMETERS

* DISTANCE INCLUDES FERRY TRAVEL

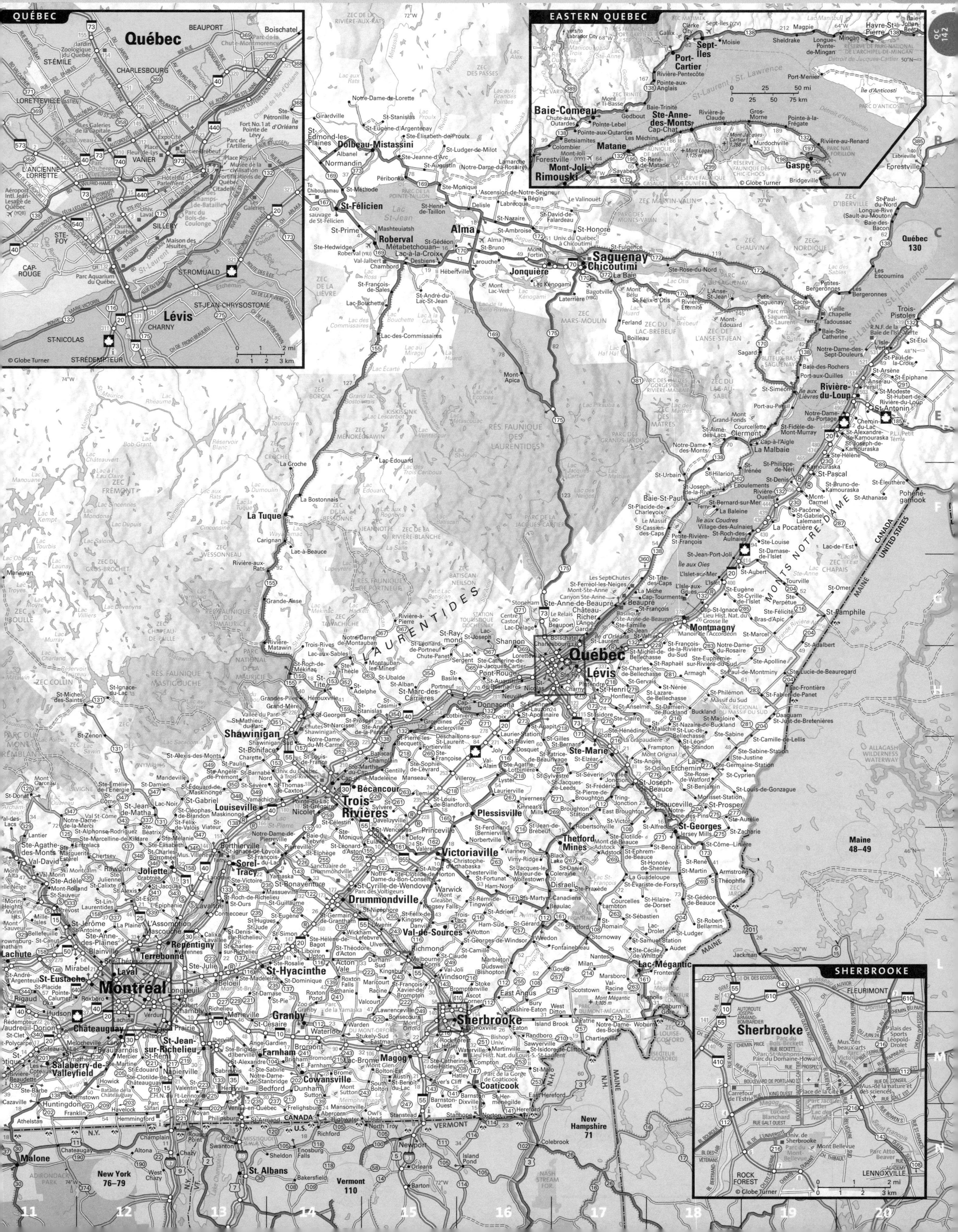

	BATHURST, NB	CAMPBELLTON, NB	CHARLOTTETOWN, PE	DIGBY, NS	EDMUNDSTON, NB	FREDERICTON, NB	GASPÉ, QC	HALIFAX, NS	LUNENBURG, NS	MIRAMICHI, NB	MONCTON, NB	NEW GLASGOW, NS	PORT HAWKESBURY, NS	RIMOUSKI, QC	RIVIÈRE-DU-LOUP, QC	SAINT JOHN, NB	ST. STEPHEN, NB	SYDNEY, NS	TRURO, NS	WOODSTOCK, NB	YARMOUTH, NS
CHARLOTTETOWN, PE	328	434		539	629	354	730	322	419	110*	258	162	110*	224*	620	749	312	417	374*	233	457	616
FREDERICTON, NB	245	351	354	669	275		647	452	549	175	192	425	539	445	395	105	123	689	363	103	746
HALIFAX, NS	452	558	322	217	727	452	854		97	382	260	151	265	744	847	410	515	415	89	555	294
SAINT JOHN, NB	350	456	312	82*	380	105	752	410	258*	280	150	383	497	550	500		105	647	321	208	176*
SYDNEY, NS	689	795	374*	632	964	689	1091	415	512	619	497	264	123	981	1084	647	752		326	792	709

DRIVING DISTANCES IN KILOMETERS
* DISTANCE INCLUDES FERRY TRAVEL

DISTANCES IN CANADA SHOWN IN KILOMETERS

© Globe Turner

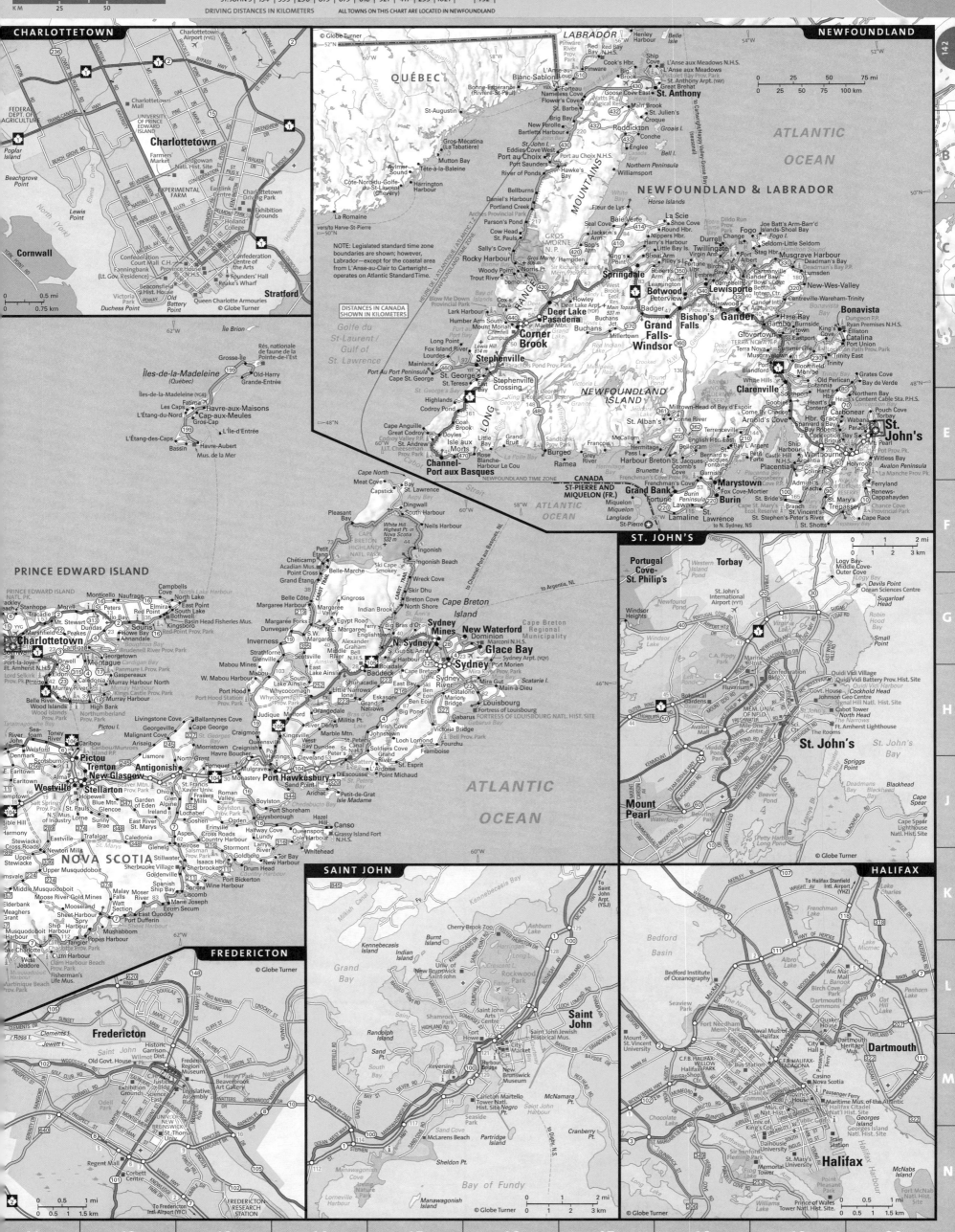

DRIVING DISTANCES IN KILOMETERS — ALL TOWNS ON THIS CHART ARE LOCATED IN NEWFOUNDLAND

	ARGENTIA	BISHOP'S FALLS	BONAVISTA	CHAN.-PT. AUX BASQUES	CORNER BROOK	DEER LAKE	GANDER	GRAND FALLS-WINDSOR	MARYSTOWN	ST. ANTHONY	ST. JOHN'S	STEPHENVILLE
ARGENTIA		363	266	845	643	588	291	381	263	991	134	702
BISHOP'S FALLS	363		307	482	280	225	72	18	384	628	393	339
CHAN.-PT. AUX BASQUES	845	482		789	202	257	554	464	866	660	875	151
CORNER BROOK	643	280	587	202		55	352	262	664	458	673	59
ST. JOHN'S	134	393	296	875	673	618	321	411	293	1021		732

DISTANCES IN CANADA SHOWN IN KILOMETERS

MI 75 150
KM 75 150

San Diego · Tijuana · Calexico · Yuma · Tecate
Rosarito · El Descanso · Mexicali · San Luis Río Colorado
El Sauzal · **Ensenada** · Constitución
Maneadero · 196 · 148 · El Faro
Punta Banda · Isla de Todos Santos · La Bufadora
Puerto Santo Tomás · San Vicente
Villa Hidalgo · Vicente Guerrero · Parque Nacional Sierra de San Pedro Mártir
Isla San Martín · San Quintín · Lázaro Cárdenas

ARIZONA
Tucson
Nogales · Naco · Douglas
Agua Prieta

NEW MEXICO
Las Cruces
Midland
El Paso · **Ciudad Juárez**

El Rosario · Punta Baja
Cataviña

BAJA CALIFORNIA

SONORA
Hermosillo
Guaymas · Empalme

CHIHUAHUA
Chihuahua
Cuauhtémoc · Delicias

COAHUILA

BAJA CALIFORNIA SUR

Ciudad Obregón
Navojoa

OCCIDENTAL
Hidalgo del Parral

SINALOA
Los Mochis
Culiacán

DURANGO
Gómez Palacio · Torreón
Durango

ZACATECAS

La Paz
Cabo San Lucas · San José del Cabo

MAZATLÁN

Mazatlán
El Castillo

OCÉANO PACÍFICO / PACIFIC OCEAN

Fresnillo
Zacatecas

NAYARIT
Tepic

Puerto Vallarta

Aguascalientes

León

MÉXICO

Cuautitlán Izcalli · Tultitlán
Villa Nicolás Romero · San Francisco Coacalco
Ciudad López Mateos · Buenavista · Ecatepec de Morelos
Santa Catarina · Acolmán · Xometla · Tepexpan
Tlalnepantla · Santa Clara · Tezoyuca · Chiconcuac
Chiautla · San Salvador Atenco
Texcoco
México
Naucalpan
Chimalhuacán
Netzahualcóyotl · San Vicente Chicoloapan
Los Reyes · Valle de Chalco · Ixtapaluca
Xochimilco · Tláhuac · Xico · Chalco
Milpa Alta

JALISCO
Guadalajara
Zapopan · Tonalá · La Piedad de Cabadas · Ocotlán
Ciudad Guzmán · Zamora de Hidalgo · Uruapan
Colima · Apatzingán
Manzanillo · Tecomán
MICHOACÁN
Lázaro Cárdenas · Zihuatanejo

OCÉANO PACÍFICO / PACIFIC OCEAN

ACAPULCO

Acapulco

OCÉANO PACÍFICO / PACIFIC OCEAN

© Globe Turner

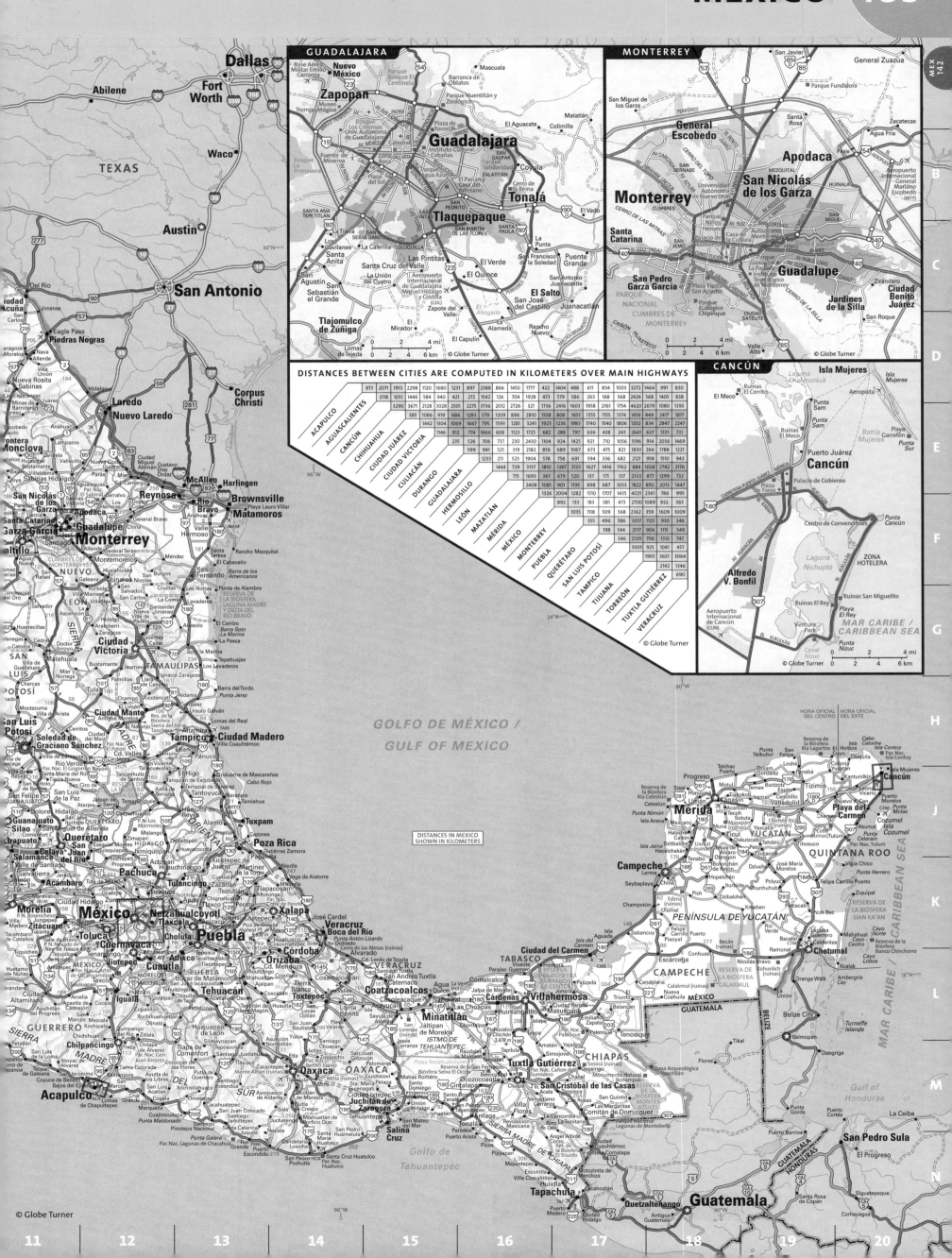

	AGUADILLA	CAGUAS	CAYEY	FAJARDO	GUAYAMA	HUMACAO	MANATÍ	MAYAGÜEZ	PONCE	SAN JUAN	UTUADO
ARECIBO	32		59	70	80	87	74	17	48	52	48	20
CAGUAS	90	59		12	36	28	17	41	97	50	17	76
MAYAGÜEZ	16	48	97	85	129	84	114	64		49	96	48
PONCE	62	52	50	39	85	37	67	52	46		67	32
SAN JUAN	80	48	17	28	34	44	33	31	96	67		67

DRIVING DISTANCES IN MILES

HORA OFICIAL DEL ATLÁNTICO
DISTANCES IN PUERTO RICO SHOWN IN KILOMETERS

OCÉANO ATLÁNTICO / ATLANTIC OCEAN

MAR CARIBE / CARIBBEAN SEA

© Globe Turner

SAN JUAN

OCÉANO ATLÁNTICO / ATLANTIC OCEAN

© Globe Turner

Alabama–California

Note: State populations are from 2020; all other figures are from 2010 or latest national census.

ALABAMA
PG. 4–5

CAPITAL
Montgomery

NICKNAME
Heart of Dixie

POPULATION
5,024,279, rank 24

AREA
51,705 sq mi, rank 29

STATEHOOD
1819, rank 22

ALASKA
PG. 6

CAPITAL
Juneau

NICKNAME
Great Land

POPULATION
733,391, rank 48

AREA
591,004 sq mi, rank 1

STATEHOOD
1959, rank 49

ARIZONA
PG. 7–9

CAPITAL
Phoenix

NICKNAME
Grand Canyon State

POPULATION
7,151,502, rank 14

AREA
114,000 sq mi, rank 6

STATEHOOD
1912, rank 48

ARKANSAS
PG. 10–11

CAPITAL
Little Rock

NICKNAME
Natural State

POPULATION
3,011,524, rank 33

AREA
53,187 sq mi, rank 27

STATEHOOD
1836, rank 25

CALIFORNIA
PG. 12–19

CAPITAL
Sacramento

NICKNAME
Golden State

POPULATION
39,538,223, rank 1

AREA
158,706 sq mi, rank 3

STATEHOOD
1850, rank 31

BIXBY BRIDGE, BIG SUR, CALIFORNIA

DENALI, DENALI NATIONAL PARK AND PRESERVE, ALASKA

FLORIDA
PG. 25-28
CAPITAL
Tallahassee
NICKNAME
Sunshine State
POPULATION
21,538,187, rank 3
AREA
58,664 sq mi, rank 22
STATEHOOD
1845, rank 27

Counties

Cities and Towns

SOUTH BEACH, MIAMI BEACH, FLORIDA

CONNECTICUT
PG. 22-23
CAPITAL
Hartford
NICKNAME
Constitution State
POPULATION
3,605,944, rank 29
AREA
5,018 sq mi, rank 48
STATEHOOD
1788, rank 5

Counties

DELAWARE
PG. 24
CAPITAL
Dover
NICKNAME
First State
POPULATION
989,948, rank 45
AREA
2,044 sq mi, rank 49
STATEHOOD
1787, rank 1

Cities and Towns

DISTRICT OF COLUMBIA
PG. 52
POPULATION
689,545
AREA
69 sq mi
BECAME CAPITAL
1800

COLORADO
PG. 19-21
CAPITAL
Denver
NICKNAME
Centennial State
POPULATION
5,773,714, rank 21
AREA
104,091 sq mi, rank 8
STATEHOOD
1876, rank 38

Counties

Cities and Towns

GEORGIA
PG. 29-31
CAPITAL
Atlanta
NICKNAME
Empire State of the South
POPULATION
10,711,908, rank 8
AREA
58,910 sq mi, rank 21
STATEHOOD
1788, rank 4

Counties

Cities and Towns

IDAHO
PG. 33
CAPITAL
Boise
NICKNAME
Gem State
POPULATION
1,839,106, rank 38
AREA
83,564 sq mi, rank 13
STATEHOOD
1890, rank 43

Counties

HAWAI'I
PG. 32
CAPITAL
Honolulu
NICKNAME
Aloha State
POPULATION
1,455,271, rank 40
AREA
6,471 sq mi, rank 47
STATEHOOD
1959, rank 50

Counties

Cities and Towns

ROCKY COAST NEAR HONOLULU, HAWAII

ILLINOIS
PG. 34-37
CAPITAL
Springfield
NICKNAME
Land of Lincoln
POPULATION
12,812,508, rank 6
AREA
56,345 sq mi, rank 24
STATEHOOD
1818, rank 21

Counties

Cities and Towns

E. Alton, 6301...P6
E. Dubuque, 1704...A4
E. Dundee, 2860...E2*
E. Galesburg, 812...F5
E. Hazel Crest, 1543...M8*
E. Moline, 21302...D5
E. Peoria, 23402...G7
E. St. Louis, 27006...Q5
Edinburg, 1078...L7
Edwardsville, 24293...P6
Effingham, 12328...N10
Elburn, 5602...C10
Eldorado, 4122...S10
Elgin, 108188...B10
Elizabeth, 761...A5
Elizabethtown, 299...U10
Elk Grove Vil., 33127...F5*
Elkville, 928...S8
Elmhurst, 44121...C11
Elmwood, 2097...G6
Elmwood Park, 24883...G6*
El Paso, 2810...G8
Elsah, 673...N5
Elwood, 2279...E11
Energy, 1146...T9
Enfield, 596...R10
Equality, 595...T10
Erie, 1602...D6
Eureka, 5295...G8
Evanston, 74486...B12
Evansville, 701...S6
Evergreen Park, 19852...K8*
Fairbury, 3757...G10
Fairfield, 5154...Q10
Fairmont City, 2635...E18†
Fairmount, 642...J12
Fairview Hts., 17078...Q6
Farmer City, 2037...J9
Farmersville, 724...M7
Farmington, 2448...G6
Findlay, 683...L9
Fisher, 1881...H10
Flanagan, 1110...F9
Flora, 5070...P10
Flossmoor, 9464...M8*
Ford Hts., 2763...N9*
Forest Lake, 1659...C4*
Forest Park, 14167...H6*
Forest View, 698...J7*
Forrest, 1220...G10
Forreston, 1446...B7
Forsyth, 3490...K9
Fox Lake, 10579...A11
Fox Lake Hills, 2591...A3*
Fox River Grove, 4854...B10
Frankfort, 17782...D11
Franklin Grove, 1021...C8
Franklin Park, 18333...G6*
Freeburg, 4354...Q6
Freeport, 25638...B7
Fulton, 3481...C5
Gages Lake, 10198...B5*
Galatia, 933...S10
Galena, 3429...A5
Galesburg, 32195...F5
Galva, 2589...E6
Gardner, 1463...E10
Geneseo, 6586...D5
Geneva, 21495...C10
Genoa, 5193...B9
Georgetown, 404...K12
Germantown, 1269...Q7
Germantown Hills, 3438...G7
Gibson City, 3407...H10
Gifford, 975...H11
Gilberts, 6879...E1*
Gillespie, 3319...N6
Gilman, 1814...G11
Girard, 2103...M6
Glasford, 1022...H6
Glen Carbon, 12934...P6
Glencoe, 8723...B12
Glendale Hts., 34208...G4*
Glen Ellyn, 27450...H4*
Glenview, 44692...E7*
Glenwood, 8969...M8*
Godfrey, 17982...P5
Golconda, 668...U10
Golden, 644...J3
Goodfield, 860...G8
Goreville, 1049...T9
Grafton, 674...N5
Grand Tower, 605...T7
Grandview, 1441...Q2
Granite City, 29849...P6
Grant Park, 1331...E12
Granville, 1427...E8
Grayslake, 20957...A11
Grayville, 1666...R11
Greenfield, 1071...M5
Green Oaks, 3866...B5*
Greenup, 1513...M11
Green Valley, 709...H7
Greenview, 778...J7
Greenville, 7000...P7
Gridley, 1432...G9
Griggsville, 1226...L4
Gurnee, 31295...A5*
Hainesville, 3597...B4*
Hamilton, 2951...H2
Hampshire, 5563...B10
Hampton, 1863...T4
Hanna City, 1225...G6
Hanover, 844...B5
Hanover Park, 37973...F3*
Hardin, 967...N4
Harrisburg, 9017...T10
Harristown, 1367...K8
Hartford, 1429...P6
Harvard, 9447...A9
Harvey, 25282...L8*
Harwood Hts., 8612...G6*
Havana, 3301...H6
Hawthorn Woods, 7663...C4*
Hazel Crest, 14100...M8*
Hebron, 1216...A10
Hennepin, 757...E8
Henry, 2464...F8
Herrin, 12501...T9
Herscher, 1591...F11
Heyworth, 2841...H9
Hickory Hills, 14049...K6*
Highland, 9919...P7
Highland Park, 29763...B12
Highwood, 5405...C6*
Hillcrest, 1326...C8
Hillsboro, 6207...N7
Hillside, 8157...H6*
Hinckley, 2070...C9
Hinsdale, 16816...J5*
Hodgkins, 1897...J6*
Hoffman Estates, 51895...F4*
Holiday Hills, 610...B2*
Homer, 1193...J11
Homer Glen, 24220...L5*
Hometown, 4349...K7*
Homewood, 19323...M8*
Hoopeston, 5351...H12
Hopedale, 865...H7
Hopkins Park, 603...F12
Hudson, 1838...G9
Huntley, 24291...B10
Hurst, 795...S8
Illiopolis, 891...K8
Ina, 2338...R9
Indian Head Park, 3809...J6*
Inverness, 7399...E4*
Irving, 495...N7
Irvington, 659...Q8
Island Lake, 8080...B11
Itasca, 8649...G4*
Jacksonville, 19446...L5
Jerome, 1656...R1
Jerseyville, 8465...N5
Johnsburg, 6337...A2*
Johnston City, 3543...S9
Joliet, 147433...D11
Jonesboro, 1821...U8
Justice, 12926...J6*
Kankakee, 27537...F11
Kansas, 787...L11
Karnak, 499...U9
Keithsburg, 609...F3
Kenilworth, 2513...E7*
Kewanee, 12916...E6
Kildeer, 3968...D4*
Kincaid, 1505...L7
Kingston, 1164...B9
Kinmundy, 796...P9
Kirkland, 1744...B9
Kirkwood, 714...G4
Knoxville, 2911...F5
Lacon, 1937...F7
Ladd, 1295...E8
La Grange, 15550...J6*
La Grange Park, 13579...H6*
La Harpe, 1235...G3
Lake Barrington, 4973...C3*
Lake Bluff, 5722...B11
Lake Forest, 19375...B11
Lake in the Hills, 28965...D1*
Lakemoor, 6017...B3*
Lake of the Woods, 2912...J10
Lake Summerset, 2048...A7
Lake Villa, 8741...A11
Lakewood, 3811...B10
Lake Zurich, 19631...B11
La Moille, 726...D8
Lanark, 1457...B6
Lansing, 28331...M9*
LaSalle, 9609...E8
Lawrenceville, 4348...P12
Lebanon, 4418...Q6
Leland, 977...D9
Leland Grove, 1503...R1
Lemont, 16000...D11
Lena, 2912...A6
Le Roy, 3560...H9
Lewistown, 2384...H5
Lexington, 2060...G9
Libertyville, 20315...B11
Lily Lake, 993...C10
Limestone, 1598...F11
Lincoln, 14504...J8
Lincolnshire, 7275...D5*
Lincolnwood, 12590...F7*
Lindenhurst, 14462...A11
Lisle, 22390...J4*
Litchfield, 6939...N7
Livingston, 858...N6
Loami, 745...L6
Lockport, 24839...D11
Lombard, 43165...H4*
Long Creek, 1328...K9
Long Grove, 8043...D5*
Long Lake, 3515...A3*
Louisville, 1139...P10
Loves Park, 23996...A8
Lovington, 1130...L10
Lynwood, 9007...D12
Lyons, 10729...J6*
Machesney Park, 23499...A8
Mackinaw, 1950...H8
Macomb, 19288...H4
Macon, 1138...L9
Madison, 3891...P5
Mahomet, 7258...J10
Malta, 1164...C9
Manhattan, 7051...E11
Manito, 1642...H6
Mansfield, 906...J10
Manteno, 9204...E12
Maple Park, 1310...C10
Marengo, 7648...B9
Marine, 960...P6
Marion, 17193...T9
Marissa, 1979...R7
Markham, 12508...M8*
Maroa, 1801...J9
Marquette Heights, 2824...D16*
Marseilles, 5094...E9
Marshall, 3933...M12
Martinsville, 1167...M12
Maryville, 7487...D19†
Mascoutah, 7483...Q6
Mason City, 2343...J7
Matherville, 723...E4
Matteson, 19009...D12
Mattoon, 18555...M10
Maywood, 24090...H6*
Mazon, 1015...E10
McCullom Lake, 1049...A2*
McHenry, 26992...A10
McLean, 830...H8
McLeansboro, 2883...R10
Melrose Park, 25411...H6*
Mendon, 953...J2
Mendota, 7372...D8
Meredosia, 1044...K4
Merrionette Park, 1900...K7*
Metamora, 3636...G8
Metropolis, 6537...V9
Midlothian, 14819...L7*
Milan, 5099...D4
Milford, 1306...G12
Milledgeville, 1032...C6
Millstadt, 4011...Q6
Minier, 1252...H8
Minonk, 2078...F8
Minooka, 10924...D10
Mokena, 18740...M6*
Moline, 43483...D5
Momence, 3310...E12
Monee, 5148...E12
Monmouth, 9444...F4
Montgomery, 18438...K1*
Monticello, 5548...J10
Morris, 13636...E10
Morrison, 4188...C6
Morrisonville, 1056...M7
Morton, 16267...G7
Morton Grove, 23270...F7*
Mound City, 588...V8
Mounds, 810...V8
Mt. Carmel, 7284...Q12
Mt. Carroll, 1717...B6
Mt. Morris, 2998...B7
Mt. Olive, 2099...N7
Mt. Prospect, 54167...E5*
Mt. Pulaski, 1566...K8
Mt. Sterling, 2025...K4
Mt. Vernon, 15277...R9
Mt. Zion, 5833...L9
Moweaqua, 1831...L8
Mulberry Grove, 634...P8
Mundelein, 31064...B11
Murphysboro, 7970...T8
Murrayville, 587...L5
Naperville, 141853...C11
Nashville, 3258...R8
Nauvoo, 1149...H2
Neoga, 1636...M10
Newark, 992...D10
New Athens, 2054...R6
New Baden, 3349...Q7
New Berlin, 1346...L6
New Boston, 683...E3
New Lenox, 24394...D11
Newman, 865...K11
Newton, 2849...N11
New Windsor, 748...E5
Niantic, 707...K8
Niles, 29803...F7*
Noble, 677...P11
Nokomis, 2256...M8
Normal, 52497...H9
Norridge, 14572...G6*
Norris City, 1275...S10
N. Aurora, 16760...C10
N. Barrington, 3047...D3*
Northbrook, 33170...B12
N. Chicago, 32574...A11
Northfield, 5420...E7*
Northlake, 12323...G5*
N. Pekin, 1573...G7
N. Riverside, 6672...H6*
Oak Brook, 7883...H5*
Oakbrook Terrace, 2134...H5*
Oak Forest, 27962...D12
Oak Grove, 396...E4
Oakland, 880...L11
Oak Lawn, 56690...C12
Oak Park, 51878...C12
Oakwood, 1595...J12
Oakwood Hills, 2083...C2*
Oblong, 1466...N11
Odell, 1046...F10
Odin, 1076...Q8
O'Fallon, 28281...Q6
Ogden, 810...J11
Oglesby, 3791...E8
Okawville, 1434...Q7
Olney, 9115...P11
Olympia Fields, 4988...N7*
Onarga, 1368...G11
Oneida, 700...F5
Oquawka, 1371...F3
Orangeville, 793...A7
Oreana, 875...K9
Oregon, 3721...B8
Orion, 1861...E5
Orland Hills, 7149...M6*
Orland Park, 56767...D11
Oswego, 30355...D10
Ottawa, 18768...E9
Palatine, 68557...B11
Palestine, 1369...N12
Palmyra, 698...M6
Palos Hts., 12515...L7*
Palos Hills, 17484...K6*
Palos Park, 4847...L6*
Pana, 5847...M8
Paris, 8837...L12
Park City, 7570...B6*
Park Forest, 21975...N8*
Park Ridge, 37480...F6*
Patoka, 584...P8
Pawnee, 2739...L7
Paw Paw, 870...D9
Paxton, 4473...H11
Payson, 1026...K3
Pearl City, 838...B6
Pecatonica, 2195...A8
Pekin, 34094...H7
Peoria, 115007...G7
Peoria Hts., 6156...G7
Peotone, 4142...E12
Percy, 970...S7
Peru, 10295...E8
Petersburg, 2260...K6
Philo, 1466...K11
Phoenix, 1964...L8*
Pierron, 600...P7
Pinckneyville, 5648...S8
Pingree Grove, 4532...B10
Piper City, 826...G11
Pistakee Highlands, 3454...A2*
Pittsfield, 4576...L4
Plainfield, 39581...D11
Plano, 10856...D10
Pleasant Hill, 966...M4
Pleasant Plains, 802...K6
Pocahontas, 784...P7
Polo, 2355...C7
Pontiac, 11931...F9
Pontoon Beach, 5836...C18†
Poplar Grove, 5023...A9
Port Barrington, 1517...C3*
Port Byron, 1647...D5
Posen, 5987...L8*
Potomac, 750...H12
Prairie du Rocher, 604...S6
Prairie Grove, 1904...C2*
Princeton, 7660...E7
Princeville, 1738...F7
Prophetstown, 2080...D6
Prospect Hts., 16256...E5*
Quincy, 40633...K2
Ramsey, 1037...N8
Rankin, 561...H11
Rantoul, 12941...H11
Rapids City, 959...T5
Raymond, 1006...M7
Red Bud, 3698...R6
Richmond, 1874...A10
Richton Park, 13646...N7*
Ridge Farm, 882...K12
Ridgway, 869...T11
Riverdale, 13549...L8*
River Forest, 11172...H6*
River Grove, 10227...G6*
Riverside, 8875...H6*
Riverton, 3455...K7
Riverwoods, 3660...D6*
Roanoke, 2065...G8
Robbins, 5337...L8*
Robinson, 7713...N12
Rochelle, 9574...C8
Rochester, 3689...L7
Rockdale, 1976...N3*
Rock Falls, 9266...C7
Rockford, 152871...B8
Rock Island, 39018...D4
Rockton, 7685...A8
Rolling Meadows, 24099...E4*
Rome, 1738...F7
Romeoville, 39680...D11
Roodhouse, 1814...M5
Roscoe, 10785...A8
Roselle, 22763...F4*
Rosemont, 4202...F6*
Roseville, 989...G4
Rosewood Hts., 4038...A18†
Rosiclare, 1160...U10
Rossville, 1331...H12
Round Lake, 18289...A4*
Round Lake Beach, 28175...A4*
Round Lake Hts., 2676...A4*
Round Lake Park, 7505...A4*
Roxana, 1542...B18†
Royalton, 1151...S8
Rushville, 3192...J4
St. Anne, 1257...F12
St. Charles, 32974...C10
St. Elmo, 1426...N9
St. Francisville, 697...Q12
St. Jacob, 1098...P7
St. Joseph, 3967...J11
Salem, 7485...Q9
Sandoval, 1274...Q8
Sandwich, 7421...D9
San Jose, 642...H7
Sauk Vil., 10506...N9*
Savanna, 3062...B6
Savoy, 7280...J10
Saybrook, 693...H10
Schaumburg, 74227...F4*
Schiller Park, 11793...G6*
Schram City, 586...N7
Seneca, 2371...E9
Sesser, 1931...S8
Shabbona, 925...C9
Shannon, 757...B7
Shawneetown, 1239...T11
Sheffield, 926...E7
Shelbyville, 4700...M9
Sheldon, 1070...G12
Sheridan, 2137...D9
Sherman, 4148...K7
Sherrard, 640...E5
Shiloh, 12651...F20†
Shipman, 624...N6
Shorewood, 15615...D11
Sidell, 617...K12
Sidney, 1233...J11
Silvis, 7479...U4
Skokie, 64784...B12
Sleepy Hollow, 3304...E1*
Smithton, 3693...Q6
Somonauk, 1893...D9
S. Barrington, 4565...E3*
S. Beloit, 7892...A8
S. Chicago Hts., 4139...N8*
S. Elgin, 21985...C10
Southern View, 1642...R2
S. Holland, 22030...M8*
S. Jacksonville, 3331...L5
S. Pekin, 1146...H7
S. Roxana, 2053...B18†
S. Wilmington, 681...E10
Sparta, 4302...R7
Springfield, 116250...K7
Spring Grove, 5778...A10
Spring Valley, 5558...E8
Stanford, 596...H8
Staunton, 5139...N6
Steeleville, 2083...S7
Steger, 9570...N8*
Sterling, 15370...C7
Stewardson, 734...M10
Stickney, 6786...J7*
Stillman Valley, 1120...B8
Stockton, 1862...A6
Stone Park, 4946...G6*
Stonington, 932...L8
Strasburg, 467...M10
Streamwood, 39858...F3*
Streator, 13710...F9
Stronghurst, 883...G3
Sugar Grove, 8997...C10
Sullivan, 4440...L10
Summit, 11054...J6*
Sumner, 3174...P12
Swansea, 13430...Q6
Sycamore, 17519...C9
Tallula, 488...K6
Tamaroa, 638...R8
Tamms, 632...V8
Tampico, 790...D6
Taylorville, 11246...L8
Teutopolis, 1530...N10
Thayer, 693...L7
Third Lake, 1182...A4*
Thomasboro, 1126...J11
Thornton, 2338...M8*
Tilden, 934...R7
Tilton, 2724...J12
Tinley Park, 56703...D12
Tiskilwa, 829...E7
Toledo, 1238...M11
Tolono, 3447...K10
Toluca, 1414...F8
Tonica, 768...E8
Toulon, 1292...F6
Tower Hill, 611...M9
Tower Lakes, 1283...C3*
Tremont, 2236...H7
Trenton, 2715...Q7
Troy, 9888...P6
Tuscola, 4480...K10
Ullin, 463...U8
University Park, 7129...N7*
Urbana, 41250...J11
Valier, 669...S8
Valmeyer, 1263...R5
Vandalia, 7042...N8
Venetian Vil., 2826...A4*
Venice, 1890...D17†
Vermont, 667...J5
Vernon Hills, 25113...C5*
Vienna, 1434...U9
Villa Grove, 2537...K11
Villa Park, 21904...H5*
Viola, 955...E4
Virden, 3425...M6
Virginia, 1611...K5
Volo, 2929...B3
Wadsworth, 3815...A11
Walnut, 1416...D7
Wamac, 1185...Q8
Wapella, 558...J9
Warren, 1428...A6
Warrensburg, 1210...K8
Warrenville, 13140...H3*
Warsaw, 1607...H2
Washburn, 1155...F8
Washington, 15134...G7
Washington Park, 4196...E18†
Wataga, 843...F5
Waterloo, 9811...R5
Waterman, 1506...C9
Watseka, 5255...G12
Watson, 754...N10
Wauconda, 13603...C3*
Waukegan, 89078...A11
Waverly, 1307...L6
Wayne, 2431...G2*
Wayne City, 1032...R10
Wenona, 1056...F8
Westchester, 16718...H6*
W. Chicago, 27086...H2*
West City, 661...S9
W. Dundee, 7331...B10
Western Sprs., 12975...J6*
Westfield, 601...M11
W. Frankfort, 8182...S9
Westmont, 24685...J5*
W. Peoria, 4458...G7
W. Salem, 897...Q11
Westville, 3202...J12
Wheaton, 52894...C11
Wheeling, 37648...B11
White Hall, 2520...M5
Williamsfield, 578...F6
Williamsville, 1476...K7
Willisville, 633...S7
Willowbrook, 8540...J5*
Willow Sprs., 5524...K6*
Wilmette, 27087...B12
Wilmington, 142...E11
Wilsonville, 586...N6
Winchester, 1593...L5
Windsor, 1187...M10
Winfield, 9080...H3*
Winnebago, 3101...B8
Winnetka, 12187...E7*
Winthrop Harbor, 6742...A11
Witt, 903...M8
Wonder Lake, 4026...A1*
Wood Dale, 13770...F5*
Woodhull, 811...E5
Woodlawn, 698...R8
Woodridge, 32971...J4*
Wood River, 10657...P6
Woodstock, 24770...A10
Worden, 1044...P6
Worth, 10789...K7*
Wyanet, 991...E7
Wyoming, 1429...F6
Yates City, 693...G6
Yorkville, 16921...D10
Zeigler, 1801...S8
Zion, 24413...A11

INDIANA

PG. 37–39

CAPITAL
Indianapolis

NICKNAME
Hoosier State

POPULATION
6,785,528, rank 17

AREA
36,185 sq mi, rank 38

STATEHOOD
1816, rank 19

Counties

Adams, 34387...F12
Allen, 355329...D11
Bartholomew, 76794...N9
Benton, 8854...F4
Blackford, 12766...G11
Boone, 56640...J7
Brown, 15242...N8
Carroll, 20155...G7
Cass, 38966...F8
Clark, 110232...R10
Clay, 26890...M5
Clinton, 33224...H7
Crawford, 10713...S7
Daviess, 31648...Q5
Dearborn, 50047...N12
Decatur, 25740...N10
DeKalb, 42223...C12
Delaware, 117671...H11
Dubois, 41889...S6
Elkhart, 197559...B9
Fayette, 24277...L11
Floyd, 74578...S9
Fountain, 17240...H4
Franklin, 23087...M12
Fulton, 20836...D8
Gibson, 33503...S3
Grant, 70061...G9
Greene, 33165...P5
Hamilton, 274569...J8
Hancock, 70002...K9
Harrison, 39364...T9
Hendricks, 145448...K7
Henry, 49462...K11
Howard, 82752...G8
Huntington, 37124...E10
Jackson, 42376...P8
Jasper, 33478...E5
Jay, 21253...G12
Jefferson, 32428...Q10
Jennings, 28525...Q10
Johnson, 139654...M8
Knox, 38440...Q4
Kosciusko, 77358...D9
LaGrange, 37128...B11
Lake, 496005...C4
LaPorte, 111467...B7
Lawrence, 46134...P7
Madison, 131636...H9
Marion, 903393...L9
Marshall, 47051...C8
Martin, 10334...Q6
Miami, 36903...F8
Monroe, 137974...P6
Montgomery, 38124...H6
Morgan, 68894...M7
Newton, 14244...E4
Noble, 47536...C11
Ohio, 6128...P12
Orange, 19840...R7
Owen, 21575...N5
Parke, 17339...K5
Perry, 19338...T6
Pike, 12845...S5
Porter, 164343...C5
Posey, 25910...T2
Pulaski, 13402...D6
Putnam, 37963...L5
Randolph, 26171...H12
Ripley, 28818...N11
Rush, 17392...L10
St. Joseph, 266931...B8
Scott, 24181...R9
Shelby, 44436...L10
Spencer, 20952...T5
Starke, 23363...D6
Steuben, 34185...B12
Sullivan, 21475...P4
Switzerland, 10613...Q12
Tippecanoe, 172780...H6
Tipton, 15936...H8
Union, 7516...M12
Vanderburgh, 179703...T3
Vermillion, 16212...K4
Vigo, 107848...M4
Wabash, 32888...E10
Warren, 8508...G4
Warrick, 59689...T4
Washington, 28262...Q8
Wayne, 68917...J12
Wells, 27636...F11
White, 24643...F6
Whitley, 33292...D11

Cities and Towns

* City indexed to pg. 36-37

Advance, 477...J7
Akron, 1167...D9
Albany, 2165...H11
Albion, 2349...C11
Alexandria, 5145...H10
Anderson, 56129...J10
Andrews, 1149...E10
Angola, 8612...B12
Arcadia, 1666...H9
Argos, 1691...D8
Ashley, 983...B12
Atlanta, 725...H9
Attica, 3245...H4
Auburn, 12731...C12
Aurora, 3750...P13
Austin, 4295...Q9
Avilla, 2401...C11
Avon, 12446...K7
Bainbridge, 746...K6
Bargersville, 4013...M8
Bass Lake, 1195...D7
Batesville, 6520...N11
Battle Ground, 1334...G6
Bedford, 13413...Q7
Beech Grove, 14192...L8
Berne, 3999...F12
Beverly Shores, 613...B5
Bicknell, 2915...Q4
Bloomfield, 2405...P6
Bloomington, 80405...N7
Bluffton, 9897...F12
Boonville, 6246...T4
Borden, 808...R9
Boswell, 778...G4
Bourbon, 1810...C8
Brazil, 7912...M5
Bremen, 4588...C8
Bright, 5693...N13
Bristol, 1602...A9
Brook, 997...E4
Brooklyn, 1598...M7
Brookston, 1554...F6
Brookville, 2596...M12
Brownsburg, 21285...K7
Brownstown, 2947...Q9
Buffalo, 692...E6
Bunker Hill, 888...F8
Burns Harbor, 1156...B5
Butler, 2684...C13
Cambridge City, 1870...K12
Camden, 611...F7
Campbellsburg, 585...R8

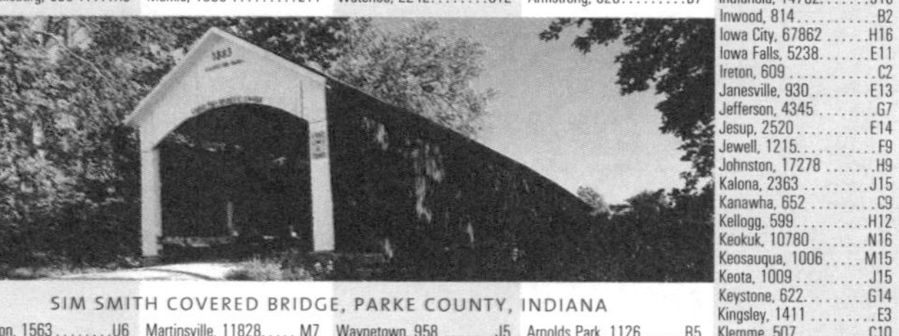

SIM SMITH COVERED BRIDGE, PARKE COUNTY, INDIANA

Cannelton, 1563...U6
Carlisle, 692...P4
Carmel, 79191...J8
Carthage, 927...K10
Cayuga, 1162...J4
Cedar Lake, 11560...C4
Centerville, 2552...K12
Chandler, 2887...T4
Charlestown, 7585...R10
Chesterfield, 2547...J10
Chesterton, 13068...B5
Churubusco, 1796...D11
Cicero, 4812...J9
Clarks Hill, 611...H6
Clarksville, 21724...S10
Clay City, 861...N5
Clayton, 972...L7
Clermont, 1356...K8
Clinton, 4893...L4
Cloverdale, 2172...M6
Colfax, 691...H6
Collegeville, 330...E5
Columbia City, 8750...D10
Columbus, 44061...N9
Connersville, 13481...L12
Converse, 1265...F9
Corydon, 3122...T8
Covington, 2645...H4
Crawfordsville, 15915...J6
Crothersville, 1591...Q9
Crown Pt., 27317...C4
Culver, 1353...D7
Cumberland, 5169...K9
Cynthiana, 545...T3
Dale, 1593...T5
Daleville, 1647...J10
Dana, 608...K4
Danville, 9001...K7
Darlington, 843...J6
Darmstadt, 1407...T3
Dayton, 1420...G6
Decatur, 9405...E12
Delphi, 2893...G6
DeMotte, 3814...D5
Dillsboro, 1327...P12
Dublin, 790...K11
Dugger, 920...P4
Dunkirk, 2362...G11
Dunlap, 6235...B9
Dyer, 16390...B4
E. Chicago, 29698...B4
Eaton, 1805...H11
Edgewood, 1913...J10
Edinburgh, 4480...M9
Elberfeld, 625...T4
Elkhart, 50949...B9
Ellettsville, 6378...N7
Elnora, 640...Q5
Elwood, 8614...H9
English, 645...S7
Etna Green, 586...C9
Evansville, 117429...U3
Fairland, 315...L9
Fairmount, 2954...G10
Fairview Park, 1386...L4
Farmersburg, 1118...N4
Farmland, 1333...H12
Ferdinand, 2157...S6
Fishers, 76794...J9
Flora, 2036...G7
Ft. Branch, 2771...S3
Fortville, 3929...J9
Ft. Wayne, 253691...D12
Fountain City, 796...J12
Fowler, 2317...F4
Francesville, 879...E6
Frankfort, 16422...H7
Franklin, 23712...M9
Frankton, 1862...H10
Freetown, 385...P8
Fremont, 2138...A12
French Lick, 1807...R7
Galena, 1818...S9
Galveston, 1311...F8
Garrett, 6286...C12
Gary, 80294...B4
Gas City, 5965...G10
Gaston, 871...H10
Geneva, 1293...F12
Georgetown, 2876...S9
Goodland, 1043...F4
Goshen, 31719...B9
Gosport, 826...M6
Grabill, 1053...D12
Grandview, 749...U5
Granger, 30465...A8
Greencastle, 10326...L6
Greendale, 4520...N13
Greenfield, 20602...K10
Greensburg, 11492...M11
Greentown, 2415...G9
Greenville, 595...S9
Greenwood, 49791...L8
Griffith, 16893...B4
Hagerstown, 1787...J12
Hamilton, 1532...B12
Hamlet, 800...C7
Hammond, 80830...B4
Hanover, 3546...Q11
Harmony, 656...M5
Hartford City, 6220...G11
Hatfield, 813...U4
Haubstadt, 1577...T3
Hebron, 3724...C5
Henryville, 1905...R10
Highland, 23727...B4
Hobart, 29059...B5
Holland, 626...S5
Homecroft, 722...L8
Hope, 2102...N10
Howe, 807...A11
Hudson, 518...B12
Huntertown, 4810...D12
Huntingburg, 6057...S5
Huntington, 17391...E10
Hymera, 801...N4
Indianapolis, 820445...K8
Ingalls, 2394...J9
Jamestown, 958...J7
Jasonville, 2222...N5
Jasper, 15038...S5
Jeffersonville, 44953...S10
Jonesboro, 1756...G10
Kendallville, 9862...C11
Kentland, 1748...F4
Kewanna, 613...E7
Kingsford Hts., 1435...B6
Kirklin, 788...H8
Knightstown, 2182...K10
Knightsville, 872...M5
Knox, 3704...C7
Kokomo, 45468...G8
Koontz Lake, 1557...C7
Kouts, 1879...C5
La Crosse, 551...C6
Ladoga, 985...K6
Lafayette, 67140...G6
La Fontaine, 875...F10
Lagrange, 2625...B11
Lakes of the Four Seasons, 7033...C5
Lake Sta., 12572...B5
Laketon, 623...E9
Lake Vil., 765...D4
Lakeville, 786...B8
Lanesville, 564...S9
Lapel, 2068...J9
La Porte, 22053...B6
Laurel, 512...M12
Lawrence, 46001...K9
Lawrenceburg, 5042...N13
Lebanon, 15792...J7
Leesburg, 555...C9
Leo-Cedarville, 3603...D12
Liberty, 2133...L12
Ligonier, 4405...B10
Linden, 759...H6
Linton, 5413...P5
Logansport, 18396...F8
Long Beach, 1179...A6
Loogootee, 2751...Q6
Lowell, 9276...C4
Lynn, 1097...J12
Lynnville, 888...T4
Lyons, 742...P5
Madison, 11967...Q11
Marengo, 828...S8
Marion, 29948...G10
Markle, 1095...E11
Martinsville, 11828...M7
Matthews, 596...G10
McCordsville, 4797...K9
Medaryville, 614...D6
Medora, 693...Q8
Mentone, 1001...D9
Meridian Hills, 1616...K8
Merrillville, 35246...B4
Mexico, 836...E8
Michigan City, 31479...A6
Middlebury, 3420...B10
Middletown, 2322...J10
Milan, 1899...N12
Milford, 1562...C9
Millersburg, 903...B10
Milltown, 818...S8
Milroy, 604...M11
Milton, 490...K12
Mishawaka, 48252...B8
Mitchell, 4350...Q7
Monon, 1777...E6
Monroe, 842...F12
Monroe City, 545...R4
Monroeville, 1235...E13
Monrovia, 1063...L7
Montezuma, 1022...K4
Monticello, 5378...F6
Montpelier, 1805...G11
Moores Hill, 597...N12
Mooresville, 9326...L7
Morgantown, 986...M8
Morocco, 1129...E4
Morristown, 1218...L10
Mt. Vernon, 6687...U2
Mulberry, 1254...H6
Muncie, 70085...H11
Munster, 23603...B4
Nappanee, 6648...C9
Nashville, 803...N8
New Albany, 36372...S9
Newburgh, 3325...U4
New Carlisle, 1861...A7
New Castle, 18114...J11
New Chicago, 2035...M12*
New Harmony, 789...T2
New Haven, 14794...D12
New Market, 636...J6
New Palestine, 2055...L9
New Paris, 1494...B9
New Pekin, 1401...R9
Newport, 515...K4
New Washington, 566...R10
New Whiteland, 5472...L8
Noblesville, 51969...J9
N. Judson, 1772...D6
N. Liberty, 1896...B7
N. Manchester, 6112...E10
N. Salem, 518...K7
N. Terre Haute, 4305...M4
N. Vernon, 6728...P10
N. Webster, 1146...C10
Oakland City, 2429...S4
Oaktown, 608...Q4
Odon, 1354...Q5
Ogden Dunes, 1110...B5
Oldenburg, 674...M12
Oolitic, 1184...P7
Orleans, 2142...Q7
Osceola, 2463...B8
Osgood, 1624...N11
Ossian, 3289...E12
Otterbein, 1262...G5
Owensville, 1284...S3
Oxford, 1162...G4
Palmyra, 930...S8
Paoli, 3677...R7
Paragon, 659...M7
Parker City, 1419...H12
Patoka, 735...S3
Pendleton, 4253...J10
Pennville, 701...G12
Peru, 11417...F9
Petersburg, 2383...R4
Pierceton, 1015...D10
Pittsboro, 2928...K7
Plainfield, 27631...L7
Plymouth, 10033...C8
Portage, 36828...B5
Porter, 4858...B5
Portland, 6223...G12
Poseyville, 1045...T3
Princes Lakes, 1312...M8
Princeton, 8644...S3
Redkey, 1353...G12
Remington, 1185...F5
Rensselaer, 5859...E5
Reynolds, 533...F6
Richmond, 36812...K13
Ridgeville, 803...H12
Rising Sun, 2304...P13
Roachdale, 926...K6
Roanoke, 1722...E11
Rochester, 6218...D8
Rockport, 2270...U5
Rockville, 2607...K5
Rocky Ripple, 606...J17*
Rolling Prairie, 582...B7
Rome City, 1361...B11
Rosedale, 725...L4
Roseland, 630...B19*
Roselawn, 4131...D4
Rossville, 1653...G7
Royal Ctr., 861...E7
Rushville, 6341...L11
Russiaville, 1094...G8
St. John, 14850...C4
St. Meinrad, 706...T6
St. Paul, 1031...M10
Salem, 6319...R8
Santa Claus, 2481...T6
Schererville, 29243...B4
Scottsburg, 6747...Q10
Seelyville, 1029...M4
Sellersburg, 6128...S10
Selma, 866...H11
Seymour, 17503...P9
Shadeland, 1610...G6
Sharpsville, 607...G8
Shelburn, 1252...N4
Shelbyville, 19191...M10
Sheridan, 2665...J8
Shirley, 830...K10
Shoals, 756...R6
Silver Lake, 915...D9
Smithville, 3184...P7
S. Bend, 101168...B8
Southport, 1712...L8
S. Whitley, 1751...D10
Speedway, 11812...K8
Spencer, 2217...N6
Spiceland, 890...K11
Staunton, 534...M5
Sullivan, 4249...P4
Summitville, 967...H10
Sunman, 1049...N12
Swayzee, 981...G9
Sweetser, 1229...G10
Syracuse, 2810...C10
Taylorsville, 919...N9
Tell City, 7272...U6
Terre Haute, 60785...M4
Thorntown, 1520...J7
Tipton, 5106...H9
Topeka, 1153...B10
Town of Pines, 708...B6
Trafalgar, 1101...M8
Trail Creek, 2052...B6
Tri-Lakes, 1421...C11
Union City, 3584...H13
Upland, 3845...G10
Utica, 776...S10
Valparaiso, 31730...C5
Van Buren, 864...F10
Veedersburg, 2180...J4
Vernon, 318...P10
Versailles, 2113...P11
Vevay, 1683...Q12
Vincennes, 18423...Q3
Wabash, 10666...E9
Wakarusa, 1758...B9
Waldron, 804...M10
Walkerton, 2144...C7
Walton, 1049...F8
Wanatah, 1048...C6
Warren, 1239...F11
Warren Park, 1480...K9
Warsaw, 13559...D9
Washington, 11509...R5
Waterloo, 2242...C12
Waynetown, 958...J5
W. Baden Sprs., 574...R7
W. College Corner, 676...L13
Westfield, 30068...J8
W. Lafayette, 29596...G6
W. Lebanon, 723...H4
Westport, 1379...N10
W. Terre Haute, 2236...M4
Westville, 5853...B6
Wheatfield, 853...D5
Whiteland, 4169...L8
Whitestown, 2867...J8
Whiting, 4997...B4
Williamsport, 1898...H4
Winamac, 2490...D7
Winchester, 4935...H12
Windfall, 708...H9
Winfield, 4383...C4
Winona Lake, 4908...D9
Winslow, 864...S4
Wolcott, 1001...F5
Wolcottville, 998...B11
Woodburn, 1520...D13
Worthington, 1463...N5
Yorktown, 9405...H10
Zanesville, 600...E11
Zionsville, 14160...J8

IOWA

PG. 40–41

CAPITAL
Des Moines

NICKNAME
Hawkeye State

POPULATION
3,190,369, rank 31

AREA
56,275 sq mi, rank 25

STATEHOOD
1846, rank 29

Counties

Adair, 7682...K7
Adams, 4029...L6
Allamakee, 14330...B16
Appanoose, 12887...M12
Audubon, 6119...H6
Benton, 26076...G14
Black Hawk, 131090...E14
Boone, 26306...G8
Bremer, 24276...D14
Buchanan, 20958...E15
Buena Vista, 20260...D5
Butler, 14867...D12
Calhoun, 9670...E7
Carroll, 20816...G6
Cass, 13956...K6
Cedar, 18499...H17
Cerro Gordo, 44151...C11
Cherokee, 12072...D4
Chickasaw, 12439...C13
Clarke, 9286...L9
Clay, 16667...C5
Clayton, 18129...D16
Clinton, 49116...G19
Crawford, 17096...G5
Dallas, 66135...H8
Davis, 8753...M13
Decatur, 8457...M9
Delaware, 17764...E16
Des Moines, 40325...L17
Dickinson, 16667...B5
Dubuque, 93653...F17
Emmet, 10302...B6
Fayette, 20880...C14
Floyd, 16303...C12
Franklin, 10680...D11
Fremont, 7441...L4
Greene, 9336...G7
Grundy, 12453...E12
Guthrie, 10954...J7
Hamilton, 15673...F9
Hancock, 11341...C9
Hardin, 17534...E11
Harrison, 14928...H3
Henry, 20145...L16
Howard, 9566...B14
Humboldt, 9815...D8
Ida, 7089...F4
Iowa, 16355...H14
Jackson, 19848...F18
Jasper, 36842...H11
Jefferson, 16843...L14
Johnson, 130882...H15
Jones, 20638...F17
Keokuk, 10511...J13
Kossuth, 15543...C8
Lee, 35862...M16
Linn, 211226...F15
Louisa, 11387...K16
Lucas, 8898...L10
Lyon, 11581...B2
Madison, 15679...K8
Mahaska, 22381...J13
Marion, 33309...J11
Marshall, 40648...G11
Mills, 15059...L4
Mitchell, 10776...B12
Monona, 9243...F2
Monroe, 7970...L12
Montgomery, 10740...L5
Muscatine, 42745...J17
O'Brien, 14398...C4
Osceola, 6462...B4
Page, 15932...M5
Palo Alto, 9421...C7
Plymouth, 24986...D2
Pocahontas, 7310...D6
Polk, 430640...H10
Pottawattamie, 93158...J4
Poweshiek, 18914...H13
Ringgold, 5131...M8
Sac, 10350...F5
Scott, 165224...H19
Shelby, 12167...J5
Sioux, 33704...C3
Story, 89542...G10
Tama, 17767...G13
Taylor, 6317...M6
Union, 12534...L8
Van Buren, 7570...M14
Wapello, 35625...L13
Warren, 46225...J10
Washington, 21704...J15
Wayne, 6403...M11
Webster, 38013...E8
Winnebago, 10866...B9
Winneshiek, 21056...B15
Woodbury, 102172...E3
Worth, 7598...B11
Wright, 13229...D9

Cities and Towns

* City indexed to pg. 35

Ackley, 1589...E11
Adair, 781...J7
Adel, 3682...H8
Afton, 845...L8
Agency, 638...L14
Akron, 1486...D1
Albert City, 699...D6
Albia, 3766...L12
Albion, 505...G12
Alburnett, 673...F15
Alden, 787...E10
Algona, 5560...C8
Allerton, 501...M10
Allison, 1029...D12
Alta, 1883...D5
Alton, 1216...C3
Altoona, 14541...H10
Amana, 442...H15
Ames, 58965...G10
Anamosa, 5533...G16
Anita, 972...J6
Ankeny, 45582...H10
Anthon, 565...E3
Aplington, 1128...E12
Armstrong, 926...B7
Arnolds Park, 1126...B5
Asbury, 4170...E18
Atkins, 1670...G15
Atlantic, 7112...J6
Audubon, 2176...H6
Aurelia, 1036...D5
Avoca, 1506...J5
Badger, 561...E8
Bancroft, 732...B8
Battle Creek, 713...F4
Baxter, 1101...H11
Bayard, 471...H7
Bedford, 1440...M6
Belle Plaine, 2534...G14
Bellevue, 2191...F19
Belmond, 2376...D10
Bertram, 294...G16
Bettendorf, 33217...J19
Blairstown, 692...G14
Bloomfield, 2640...M13
Blue Grass, 1452...J18
Bondurant, 3860...H10
Boone, 12661...G9
Boyden, 707...B3
Brighton, 652...K15
Britt, 2069...C9
Brooklyn, 1468...H13
Buffalo, 1270...J18
Buffalo Ctr., 905...B9
Burlington, 25663...M17
Burt, 533...B8
Calmar, 978...C15
Camanche, 4448...H19
Cambridge, 827...G10
Carlisle, 3876...J10
Carroll, 10103...G6
Carson, 812...K4
Carter Lake, 3785...K3
Cascade, 2159...F17
Cedar Falls, 39260...E13
Cedar Rapids, 126326...G15
Center Pt., 2421...F15
Centerville, 5528...M12
Central City, 1257...F16
Chariton, 4321...L11
Charles City, 7652...C12
Charter Oak, 502...G4
Cherokee, 5253...D4
Clarence, 974...G17
Clarinda, 5572...M5
Clarion, 2850...D9
Clarksville, 1439...D12
Clear Lake, 7777...C10
Clermont, 632...C15
Clinton, 26885...G19
Clive, 15447...B17
Coggon, 658...F16
Colfax, 2093...H11
Colo, 876...G11
Columbus Jct., 1899...K16
Conrad, 1108...F12
Coon Rapids, 1305...H7
Coralville, 18907...H16
Corning, 1635...L6
Correctionville, 821...E4
Corydon, 1585...M10
Council Bluffs, 62230...K3
Crescent, 617...J3
Cresco, 3868...B14
Creston, 7834...L7
Dakota City, 843...D8
Dallas Ctr., 1623...H9
Danville, 934...L16
Davenport, 99685...J18
Dayton, 837...F8
Decorah, 8127...B15
Denison, 8298...G5
Denver, 1780...D13
Des Moines, 203433...J10
De Soto, 1050...J8
De Witt, 5322...H19
Dexter, 611...J8
Dike, 1209...E13
Donnellson, 912...M16
Doon, 577...B2
Dows, 538...E10
Dubuque, 57637...E18
Dumont, 637...D12
Dunkerton, 852...E14
Dunlap, 1042...H4
Durant, 1832...H17
Dyersville, 4058...E17
Dysart, 1379...F13
Eagle Grove, 3583...D9
Earlham, 1450...J8
Earlville, 812...E16
Early, 557...E5
Eddyville, 1024...K13
Edgewood, 864...E16
Eldon, 927...L14
Eldora, 2732...F11
Eldridge, 5651...H18
Elgin, 683...C15
Elkader, 1273...D16
Elk Horn, 662...J5
Elk Run Hts., 1117...E14
Ellsworth, 531...F10
Elma, 546...B13
Ely, 1776...G16
Emmetsburg, 3904...C7
Epworth, 1860...E17
Essex, 798...L5
Estherville, 6360...B6
Evansdale, 4751...E14
Everly, 603...C5
Exira, 840...J6
Fairbank, 1113...E14
Fairfax, 2123...G15
Fairfield, 9464...L15
Farley, 1537...E17
Farmington, 664...M15
Fayette, 1338...D15
Fonda, 631...E6
Fontanelle, 672...K7
Forest City, 4151...B10
Ft. Dodge, 25206...E8
Ft. Madison, 11051...M16
Fredericksburg, 931...C14
Fremont, 743...K13
Fruitland, 977...J17
Garnavillo, 745...D16
Garner, 3129...C10
Garwin, 527...G12
George, 1080...B3
Gilbert, 1082...G9
Gilbertville, 712...E14
Gilman, 509...G12
Gilmore City, 504...D7
Gladbrook, 945...F12
Glenwood, 5269...L3
Glidden, 1146...G6
Goldfield, 635...D9
Gowrie, 1037...F8
Graettinger, 844...B6
Grand Jct., 824...G8
Grand Mound, 642...H18
Grandview, 556...K17
Granger, 1244...H9
Greene, 1130...D12
Greenfield, 1982...K7
Grimes, 8246...H9
Grinnell, 9218...H12
Griswold, 1036...K5
Grundy Ctr., 2706...F12
Guthrie Ctr., 1569...H7
Guttenberg, 1919...D17
Hamburg, 1187...M4
Hampton, 4461...D11
Harlan, 5106...H5
Hartford, 771...J10
Hartley, 1672...C4
Hawarden, 2546...C2
Hazleton, 823...E15
Hedrick, 764...K14
Hiawatha, 7024...G15
Hills, 703...J16
Hinton, 928...E2
Holstein, 1396...E4
Hopkinton, 628...F16
Hospers, 698...C3
Hubbard, 845...F11
Hudson, 2282...E13
Hull, 2175...B3
Humboldt, 4690...D8
Humeston, 494...L10
Huxley, 3317...G10
Ida Grove, 2142...F4
Independence, 5966...E15
Indianola, 14782...J10
Inwood, 814...B2
Iowa City, 67862...H16
Iowa Falls, 5238...E11
Ireton, 609...C2
Janesville, 930...E13
Jefferson, 4345...G7
Jesup, 2520...E14
Jewell, 1215...F9
Johnston, 17278...H9
Kalona, 2363...J15
Kanawha, 652...C9
Kellogg, 599...H12
Keokuk, 10780...N16
Keosauqua, 1006...M15
Keota, 1009...J15
Keystone, 622...G14
Kingsley, 1411...E3
Klemme, 507...C10
Knoxville, 7313...K11
Lake City, 1727...F6
Lake Mills, 2100...B10
Lake Park, 1105...A5
Lake View, 1142...F6
Lamoni, 2324...M9
Lansing, 999...B16
La Porte City, 2285...F14
Larchwood, 866...A2
Latimer, 507...D10
Laurens, 1258...D6
Lawton, 908...E2
Le Claire, 3765...H19
Le Grand, 938...G12
Le Mars, 9826...D2
Lenox, 1407...L7
Leon, 1977...M9
Lisbon, 2152...G16
Logan, 1534...H3
Lone Tree, 1300...J16
Long Grove, 808...H18
Lovilia, 538...K12
Lowden, 789...G17
Madrid, 2543...H9
Malvern, 1142...L4
Manchester, 5179...E16
Manilla, 776...G5
Manly, 1323...B11
Manning, 1500...G6
Manson, 1690...E7
Mapleton, 1224...F3
Maquoketa, 6141...G18
Marcus, 1117...D4
Marengo, 2528...H14
Marion, 34768...G15
Marshalltown, 27552...G12
Mason City, 28079...C11
Maxwell, 920...G10
McGregor, 871...C17
Mechanicsville, 1146...G16
Mediapolis, 1560...L17
Melbourne, 830...G11
Melcher-Dallas, 1288...K11
Merrill, 755...D2
Middletown, 318...L17
Milford, 2898...B5
Milo, 775...K10
Milton, 443...M14
Minden, 599...J4
Missouri Valley, 2838...J3
Mitchellville, 2254...H10
Monona, 1549...C16
Monroe, 1830...J11
Montezuma, 1462...J13
Monticello, 3796...F17
Montrose, 898...N16
Moravia, 665...L12
Morning Sun, 836...K17
Moulton, 605...M13
Mt. Ayr, 1691...M8
Mt. Pleasant, 8668...L16
Mt. Vernon, 4506...G16
Moville, 1618...E3
Murray, 756...L9
Muscatine, 22886...J17
Mystic, 425...M12
Nashua, 1663...C13
Neola, 842...J4
Nevada, 6798...G10
New Albin, 522...A16
Newell, 876...E6
Newhall, 875...G14
New Hampton, 3571...C13
New Hartford, 516...E13
New London, 1897...L16
New Sharon, 1293...J13
Newton, 15254...H11
Nora Sprs., 1431...C12
N. English, 1041...J14
N. Liberty, 13374...H16
Northwood, 1989...B11
Norwalk, 8945...J9
Norway, 545...G15
Oakland, 1527...K5
Ocheyedan, 490...B4
Odebolt, 1013...F5
Oelwein, 6415...D15
Ogden, 2044...G8
Okoboji, 807...B5
Olin, 698...G17
Onawa, 2998...G3
Orange City, 6004...C3
Orleans, 608...A6
Osage, 3619...B12
Osceola, 4929...L9
Oskaloosa, 11463...K13
Ossian, 845...C15
Otho, 542...E8
Ottumwa, 25023...L13
Oxford, 807...H15
Oxford Jct., 496...G17
Palo, 1026...G15
Panora, 1124...H7
Parkersburg, 1870...E12
Park View, 2389...H19
Paullina, 1056...C4
Pella, 10352...J12
Peosta, 1377...E18
Perry, 7702...H8
Pleasant Hill, 8785...J10
Pleasantville, 1694...J11
Pocahontas, 1789...D7
Polk City, 3418...H9
Pomeroy, 662...E7
Postville, 2227...C15
Prairie City, 1680...J11
Preston, 1012...G19
Primghar, 909...C4
Princeton, 886...H19
Quasqueton, 554...E15
Radcliffe, 545...F10
Readlyn, 808...D14
Redfield, 835...J8
Red Oak, 5742...L5
Reinbeck, 1664...F13
Remsen, 1663...D3
Riceville, 785...B13
Richland, 584...K14
Riverdale, 405...U3*
Riverside, 993...J16
Robins, 3142...G15
Rockford, 860...C12
Rock Rapids, 2549...A3
Rock Valley, 3354...B2
Rockwell, 1039...C11
Rockwell City, 1709...F7
Roland, 1284...F10
Rolfe, 584...D7
Russell, 554...L11
Ruthven, 737...C6
Sabula, 576...G19
Sac City, 2220...E6
St. Ansgar, 1107...B12
St. Charles, 653...K9
Sanborn, 1404...B4
Saylorville, 3301...H10
Schaller, 772...E5
Schleswig, 882...F4
Scranton, 557...G7
Sergeant Bluff, 4227...E2
Seymour, 701...M11
Sheffield, 1172...D11
Shelby, 641...J4
Sheldon, 5188...B3
Shell Rock, 1296...D13
Shellsburg, 983...G15
Shenandoah, 5150...M5
Sibley, 2798...B4
Sidney, 1138...M4
Sigourney, 2059...K14
Sioux Ctr., 7048...C3
Sioux City, 82684...E2
Sioux Rapids, 775...D5
Slater, 1489...H9
Sloan, 973...F2
Solon, 2037...H16
Spencer, 11233...C5
Spirit Lake, 4840...B6
Springville, 1074...G16
Stanton, 689...L5
Stanwood, 684...G17
State Ctr., 1468...G11
Storm Lake, 10600...E5
Story City, 3431...F10
Stratford, 743...F9
Strawberry Pt., 1279...D16
Stuart, 1648...J8
Sully, 821...J12
Sumner, 2028...D14
Sutherland, 649...C4
Swea City, 536...B8
Swisher, 879...H15
Tabor, 1040...L4
Tama, 2877...G13
Thompson, 502...B9
Tiffin, 1947...H15
Tipton, 3221...H17
Titonka, 476...B8
Toledo, 2341...G13
Traer, 1703...F13
Treynor, 919...K4
Tripoli, 1313...D14
Underwood, 917...J4
University Park, 487...K13
Urbana, 1458...F15
Urbandale, 39463...H9
Van Horne, 682...G14
Van Meter, 1016...J9
Ventura, 717...C10
Victor, 893...H14
Villisca, 1252...L6
Vinton, 5257...F14
Walcott, 1629...H18
Walford, 1463...G15
Walker, 791...F15
Wall Lake, 819...F5
Walnut, 785...J5
Wapello, 2067...K17
Washburn, 876...E14
Washington, 7266...K15
Waterloo, 68406...E13
Waukee, 13790...J9
Waukon, 3897...B16
Waverly, 9874...D13
Wayland, 966...K15
Webster City, 8070...E9
Wellman, 1408...J15
Wellsburg, 707...E12
W. Bend, 785...C7
W. Branch, 2322...H16
W. Burlington, 2968...L17
W. Des Moines, 56609...J9
W. Liberty, 3736...J16
W. Point, 966...M16
W. Union, 2486...C15
What Cheer, 646...J13
Wheatland, 764...H18
Whiting, 762...F2
Whittemore, 504...C7
Williamsburg, 3068...H14
Wilton, 2802...J17
Windsor Hts., 4860...B18
Winfield, 1134...K16
Winterset, 5190...K8
Winthrop, 850...E15
Woodbine, 1459...H4
Woodward, 1024...H9
Wyoming, 515...G17
Zearing, 554...F11

KANSAS

PG. 42–43

CAPITAL
Topeka

NICKNAME
Sunflower State

POPULATION
2,937,880, rank 35

AREA
82,277 sq mi, rank 14

STATEHOOD
1861, rank 34

Counties

Allen, 13371...H17
Anderson, 8102...F17
Atchison, 16924...C17
Barber, 4861...J9
Barton, 27674...F9
Bourbon, 15173...G18
Brown, 9984...B16
Butler, 65880...H14
Chase, 2790...F14
Chautauqua, 3669...K15
Cherokee, 21603...K18
Cheyenne, 2726...B2
Clark, 2215...J6
Clay, 8535...C13
Cloud, 9533...C12
Coffey, 8601...F16
Comanche, 1891...K8
Cowley, 36311...J14
Crawford, 39134...H18
Decatur, 2961...B5
Dickinson, 19754...E13
Doniphan, 7945...B17
Douglas, 110826...E17
Edwards, 3037...H8
Elk, 2882...J15
Ellis, 28452...D7
Ellsworth, 6497...E10
Finney, 36776...G4
Ford, 33848...H6
Franklin, 25992...F17
Geary, 34362...D14
Gove, 2695...E5
Graham, 2597...C6
Grant, 7829...H3
Gray, 6006...H5
Greeley, 1247...E2
Greenwood, 6689...G15
Hamilton, 2690...G2
Harper, 6034...K11
Harvey, 34684...G12
Haskell, 4256...H4
Hodgeman, 1916...G6
Jackson, 13462...C16
Jefferson, 19126...C17
Jewell, 3077...B10
Johnson, 544179...E18
Kearny, 3977...G3
Kingman, 7858...H11
Kiowa, 2553...J8
Labette, 21607...J17
Lane, 1750...F5
Leavenworth, 76227...D17
Lincoln, 3241...D10
Linn, 9656...F18
Logan, 2756...E3
Lyon, 33690...F15
McPherson, 29180...F11
Marion, 12660...G13
Marshall, 10117...B14
Meade, 4575...J5
Miami, 32787...E18
Mitchell, 6373...C10
Montgomery, 35471...J16
Morris, 5923...E14
Morton, 3233...J2
Nemaha, 10178...B15
Neosho, 16512...H17
Ness, 3107...F7
Norton, 5671...B7
Osage, 16295...F16
Osborne, 3858...C9
Ottawa, 6091...D12
Pawnee, 6973...F8
Phillips, 5642...B8
Pottawatomie, 21604...C15
Pratt, 9656...H9
Rawlins, 2519...B4
Reno, 64511...G11
Republic, 4980...B12
Rice, 10083...F10
Riley, 71115...C13
Rooks, 5181...C8
Rush, 3307...E8
Russell, 6970...D9
Saline, 55606...E11
Scott, 4936...F4
Sedgwick, 498365...H12
Seward, 22952...J4
Shawnee, 177934...D16
Sheridan, 2556...C5
Sherman, 6010...C2
Smith, 3853...B9
Stafford, 4437...G9
Stanton, 2235...H2
Stevens, 5724...J3
Sumner, 24132...K12
Thomas, 7900...C4
Trego, 3001...E6
Wabaunsee, 7053...D15
Wallace, 1485...D2
Washington, 5799...B12
Wichita, 2234...F3
Wilson, 9409...H16
Woodson, 3309...G16
Wyandotte, 157505...D18

Cities and Towns

* City indexed to pg. 63

Abilene, 6844...E13
Alma, 832...D15
Almena, 408...B7
Altamont, 1080...J17
Alta Vista, 444...E14
Altoona, 414...J16
Americus, 894...F15
Andale, 928...H12
Andover, 11791...H13
Anthony, 2269...K11
Argonia, 501...J11
Arkansas City, 12415...K13
Arlington, 473...H10
Arma, 1481...H18
Ashland, 867...K7
Assaria, 413...E12
Atchison, 11021...C17
Attica, 626...J10
Atwood, 1194...B4
Auburn, 1227...D16
Augusta, 9274...H13
Axtell, 406...B15
Baldwin City, 4515...E17
Basehor, 4613...D18
Baxter Sprs., 4238...K18
Bel Aire, 6769...L19
Belle Plaine, 1681...J13
Belleville, 1991...B12
Beloit, 3835...C11
Bennington, 672...D12
Benton, 880...H13
Bird City, 447...B3
Blue Rapids, 1019...B14
Bonner Sprs., 7314...D18
Bucklin, 794...J7
Buhler, 1327...G11
Burden, 535...J14
Burlingame, 934...E16
Burlington, 2674...G16
Burrton, 901...G12
Caldwell, 1068...K12
Caney, 2203...K16
Canton, 748...F12
Carbondale, 1437...E16
Cawker City, 469...C10
Cedar Vale, 579...K14
Centralia, 512...B15
Chanute, 9119...H17
Chapman, 1393...D13
Chase, 477...F10
Cheney, 2094...H11
Cherokee, 714...J18
Cherryvale, 2367...J17
Chetopa, 1125...K18
Cimarron, 2184...H5
Claflin, 645...F9
Clay Ctr., 4334...C13
Clearwater, 2481...J12
Clifton, 554...C13
Clyde, 716...C12
Coffeyville, 10295...K16
Colby, 5387...C4
Coldwater, 828...J8
Columbus, 3312...K18
Colwich, 1327...H12
Concordia, 5395...C12
Conway Sprs., 1272...J12
Cottonwood Falls, 903...F14
Council Grove, 2182...E14
Cunningham, 454...H10
Deerfield, 700...G4
Delphos, 359...C11
Derby, 22158...J13
De Soto, 5720...D18
Dighton, 1038...F5
Dodge City, 27340...H6
Douglass, 1700...J13
Downs, 900...C9
Eastborough, 773...M19
Edgerton, 1671...E18
Edna, 442...K17
Edwardsville, 4340...F1*
Effingham, 546...C17
El Dorado, 13021...H14
Elkhart, 2205...K2
Ellinwood, 2131...F9
Ellis, 2062...D7
Ellsworth, 3120...E10
Elwood, 1224...B18
Emporia, 24916...F15
Enterprise, 855...E13
Erie, 1150...H17
Eskridge, 534...E15
Eudora, 6136...D17
Eureka, 2633...H15
Fairway, 3882...G3*
Florence, 465...G13
Ft. Dodge, 165...H6
Ft. Scott, 8087...H18
Fowler, 590...J6
Frankfort, 726...B14
Fredonia, 2482...J16
Frontenac, 3437...J19
Galena, 3085...K19
Galva, 870...F12
Garden City, 26658...G4
Garden Plain, 849...H12
Gardner, 19123...E18
Garnett, 3415...F17
Gas, 564...G17
Girard, 2789...H18
Glasco, 498...C11
Glen Elder, 445...C10
Goddard, 4344...H12
Goessel, 539...F12
Goodland, 4489...C2
Gove, 80...D5

KENTUCKY HORSE FARM

Grandview Plaza, 1560...D14
Great Bend, 15995...F9
Greensburg, 777...H8
Halstead, 2085...G12
Hanover, 682...B13
Harper, 1473...J11
Hartford, 371...F16
Haven, 1237...H11
Haviland, 701...H8
Hays, 20510...E8
Haysville, 10826...H12
Herington, 2526...E13
Hesston, 3709...G12
Hiawatha, 3172...B16
Highland, 1012...B17
Hill City, 1474...C7
Hillsboro, 2993...F13
Hoisington, 2706...F9
Holcomb, 2094...G4
Holton, 3329...C16
Holyrood, 447...F10
Horton, 1776...B16
Howard, 687...J15
Hoxie, 1201...C5
Hoyt, 669...C16
Hugoton, 3904...K3
Humboldt, 1953...H17
Hutchinson, 42080...G11
Independence, 9483...J16
Inman, 1377...G11
Iola, 5704...G17
Jetmore, 867...G6
Jewell, 432...B10
Johnson City, 1495...H2
Junction City, 23353...D14
Kanopolis, 492...E10
Kansas City, 145786...D18
Kechi, 1909...H13
Kensington, 473...B8
Kingman, 3177...H11
Kinsley, 1457...G7
Kiowa, 1026...K10
Kismet, 459...K5
La Crosse, 1342...F8
La Cygne, 1149...F18
La Harpe, 578...G17
Lake Quivira, 906...F1*
Lakin, 2216...G3
Lansing, 11265...C18
Larned, 4054...G8
Lawrence, 87643...D17
Leavenworth, 35251...C18
Leawood, 31867...J3*
Lebo, 940...F16
Lecompton, 625...D17
Lenexa, 48190...D18
Leon, 704...H14
Leoti, 1534...F3
Le Roy, 561...G16
Lewis, 451...G8
Liberal, 20525...K4
Lincoln, 1297...D10
Lindsborg, 3458...F12
Linn, 410...B13
Linn Valley, 804...F18
Little River, 557...F11
Logan, 589...B7
Louisburg, 4315...E18
Lucas, 393...D9
Lyndon, 1052...E16
Lyons, 3739...F10
Macksville, 549...G8
Madison, 701...G15
Maize, 3420...H12
Manhattan, 52281...D14
Mankato, 869...B10
Maple Hill, 620...D15
Marion, 1927...F13
Marquette, 641...F11
Marysville, 3294...B14
McCune, 405...J18
McLouth, 880...D17
McPherson, 13155...F12
Meade, 1721...J5
Medicine Lodge, 2009...J9
Melvern, 385...F16
Meriden, 813...D16
Merriam, 11003...G3*
Milford, 530...D13
Miltonvale, 539...C12
Minneapolis, 2032...D12
Minneola, 745...J6
Mission, 9323...G3*
Mission Hills, 3498...G4*
Moline, 371...J15
Montezuma, 966...H5
Moran, 558...G17
Mound City, 694...G18
Moundridge, 1737...G12
Mound Valley, 407...J17
Mt. Hope, 813...H12
Mulberry, 520...H19
Mulvane, 6111...J13
Neodesha, 2486...J16
Ness City, 1449...F6
New Strawn, 394...F16
Newton, 19132...G12
Nickerson, 1070...G11
N. Newton, 1759...G12
Norton, 2928...B6
Nortonville, 637...C17
Norwich, 491...J11
Oaklawn, 3276...N19
Oakley, 2045...D4
Oberlin, 1788...B5
Ogden, 2087...D14
Olathe, 125872...D18
Olpe, 546...F15
Onaga, 702...C15
Osage City, 2943...E16
Osawatomie, 4447...F18
Osborne, 1431...C9
Oskaloosa, 1113...D17
Oswego, 1829...K18
Ottawa, 12649...E17
Overbrook, 1058...E16
Overland Park, 173372...D18
Oxford, 1049...J13
Ozawkie, 645...D17
Paola, 5602...E18
Park City, 7297...H13
Parsons, 10500...J17
Peabody, 1210...G13
Perry, 929...D17
Phillipsburg, 2581...B8
Pittsburg, 20233...J18
Plains, 1146...J5
Plainville, 1903...D8
Pleasanton, 1216...G18
Pomona, 832...E17
Potwin, 449...G13
Prairie Vil., 21447...G3*
Pratt, 6835...H9
Pretty Prairie, 680...H11
Protection, 514...K7
Quenemo, 388...E17
Quinter, 918...D6
Richmond, 464...F17
Riley, 939...C14
Riverton, 929...K19
Roeland Park, 6731...F3*
Rolla, 442...K2
Rose Hill, 3931...H13
Rossville, 1151...D16
Russell, 4506...E9
Sabetha, 2571...B16
St. Francis, 1329...B2
St. George, 639...D14
St. John, 1295...G9
St. Marys, 2627...D15
St. Paul, 629...J17
Salina, 47707...E12
Satanta, 1133...J4
Scammon, 482...J18
Scandia, 372...B11
Scott City, 3816...F4
Scranton, 710...E16
Sedan, 1124...K15
Sedgwick, 1695...G12
Seneca, 1991...B15
Sharon Sprs., 748...D2
Shawnee, 62209...D18
Silver Lake, 1439...D16
Smith Ctr., 1665...B9
Solomon, 1095...E12
S. Hutchinson, 2457...G11
Spearville, 773...H7
Spring Hill, 5437...E18
Stafford, 1042...G9
Sterling, 2328...G10
Stockton, 1329...C8
Strong City, 485...F14
Sublette, 1453...J4
Syracuse, 1812...G2
Thayer, 497...J17
Tonganoxie, 4996...D17
Topeka, 127473...D16
Towanda, 1450...H13
Tribune, 741...F2
Troy, 1010...B17
Turon, 387...H10
Udall, 746...J13
Ulysses, 6161...H3
Valley Ctr., 6822...H12
Valley Falls, 1192...C17
Victoria, 1214...E8
WaKeeney, 1862...D6
Wakefield, 980...D13
Wamego, 4372...D15
Washington, 1131...B13
Waterville, 680...B14
Wathena, 1364...B18
Waverly, 592...F16
Weir, 686...J18
Wellington, 8172...J12
Wellsville, 1857...E17
Westmoreland, 778...C14
Westwood, 1506...G4*
White City, 618...E14
Whitewater, 718...G13
Wichita, 382368...H12
Wilson, 781...E10
Winchester, 551...C17
Winfield, 12301...J13
Yates Ctr., 1417...G16

KENTUCKY

PG. 44–45

CAPITAL
Frankfort

NICKNAME
Bluegrass State

POPULATION
4,505,836, rank 26

AREA
40,409 sq mi, rank 37

STATEHOOD
1792, rank 15

Counties

Adair, 18656...L10
Allen, 19956...M7
Anderson, 21421...G11
Ballard, 8249...D3
Barren, 42173...M7
Bath, 11591...G14
Bell, 28691...M15
Boone, 118811...D12
Bourbon, 19985...F13
Boyd, 49542...F18
Boyle, 28432...J11
Bracken, 8488...E14
Breathitt, 13878...J16
Breckinridge, 20059...H6
Bullitt, 74319...H9
Butler, 12690...L6
Caldwell, 12984...K2
Calloway, 37191...N1
Campbell, 90336...D13
Carlisle, 5104...E3
Carroll, 10811...E11
Carter, 27720...F17
Casey, 15955...K11
Christian, 73955...L3
Clark, 35613...H14
Clay, 21730...L15
Clinton, 10272...M11
Crittenden, 9315...K1
Cumberland, 6856...M10
Daviess, 96656...J4
Edmonson, 12161...K7
Elliott, 7852...G17
Estill, 14672...H14
Fayette, 295803...G13
Fleming, 14348...F15
Floyd, 39451...J18
Franklin, 49285...F11
Fulton, 6813...G2
Gallatin, 8589...E11
Garrard, 16912...J12
Grant, 24662...E12
Graves, 37121...F4
Grayson, 25746...K6
Green, 11258...K9
Greenup, 36910...E17
Hancock, 8565...H5
Hardin, 105543...H8
Harlan, 29278...M17
Harrison, 18846...F13
Hart, 18199...K8
Henderson, 46250...J3
Henry, 15416...F11
Hickman, 4902...F3
Hopkins, 46920...K3
Jackson, 13494...J14
Jefferson, 741096...G9
Jessamine, 48586...H12
Johnson, 23356...H18
Kenton, 159720...D12
Knott, 16346...K17
Knox, 31883...M15
Larue, 14193...J9
Laurel, 58849...L13
Lawrence, 15860...G18
Lee, 7887...J15
Leslie, 11310...L16
Letcher, 24519...L17
Lewis, 13870...E16
Lincoln, 24742...K12
Livingston, 9519...K1
Logan, 26835...M5
Lyon, 8314...L1
McCracken, 65565...E3
McCreary, 18306...M13
McLean, 9531...J4
Madison, 82916...H13
Magoffin, 13333...H17
Marion, 19820...J10
Marshall, 31448...M1
Martin, 12929...H19
Mason, 17490...E15
Meade, 28602...H7
Menifee, 6306...H15
Mercer, 21331...H11
Metcalfe, 10099...L9
Monroe, 10963...M9
Montgomery, 26499...G14
Morgan, 13923...H17
Muhlenberg, 31499...L4
Nelson, 43437...H10
Nicholas, 7135...F14
Ohio, 23842...J5
Oldham, 60316...F9
Owen, 10841...E11
Owsley, 4755...K15
Pendleton, 14877...E13
Perry, 28712...L17
Pike, 65024...J19
Powell, 12613...H15
Pulaski, 63063...L13
Robertson, 2282...E14
Rockcastle, 17056...K13
Rowan, 23333...F16
Russell, 17565...L11
Scott, 47173...F12
Shelby, 42074...F10
Simpson, 17327...M6
Spencer, 17061...H10
Taylor, 24512...K10
Todd, 12460...M4
Trigg, 14339...M2
Trimble, 8809...E10
Union, 15007...J2
Warren, 113792...L6
Washington, 11717...J10
Wayne, 20813...M12
Webster, 13621...J3
Whitley, 35637...M13
Wolfe, 7355...H15
Woodford, 24939...G12

Cities and Towns

* City indexed to pg. 88

Adairville, 852...N5
Albany, 2033...N11
Alexandria, 8477...D13
Anchorage, 2348...A10
Annville, 1095...K14
Arlington, 324...F3
Ashland, 21684...E18
Auburn, 1340...M6
Audubon Park, 1473...C7
Augusta, 1190...D14
Bancroft, 494...A9
Barbourmeade, 1218...A9
Barbourville, 3165...M15
Bardstown, 11700...H9
Bardwell, 723...E3
Barlow, 675...E2
Beattyville, 1307...J15
Beaver Dam, 3409...K5
Bedford, 599...E10
Beechwood Vil., 1324...B9
Bellefonte, 888...E18
Bellemeade, 865...B9
Bellevue, 5955...C13
Benham, 500...L17
Benton, 4349...E5
Berea, 13561...J13
Bloomfield, 838...H10
Blue Ridge Manor, 767...B10
Booneville, 81...J15
Bowling Green, 58067...L6
Brandenburg, 2643...G7
Briarwood, 435...A9
Brodhead, 1211...K13
Bromley, 763...X2*
Brooks, 2401...G9
Brooksville, 642...E14
Brownsville, 836...L7
Buckner, 5837...F9
Buffalo, 498...J9
Burgin, 965...J12
Burkesville, 1521...M10
Burlington, 15926...C12
Burnside, 611...L12
Butler, 612...D13
Cadiz, 2558...M2
Calhoun, 763...J4
Calvert City, 2566...E5
Camargo, 1081...G14
Campbellsburg, 813...E10
Campbellsville, 9108...K10
Campton, 441...H16
Caneyville, 608...K6
Cannonsburg, 856...F18
Carlisle, 2010...F14
Carrollton, 3938...E10
Catlettsburg, 1856...F18
Cave City, 2240...L8
Cawood, 731...M17
Cecilia, 572...J8
Centertown, 423...K5
Central City, 5978...K4
Chaplin, 418...H10
Clarkson, 875...J7
Claryville, 2355...D13
Clay, 1181...J2
Clay City, 1077...H14
Clinton, 1388...F3
Cloverport, 1152...H6
Coal Run Vil., 1706...J19
Cold Spr., 5912...Y3*
Columbia, 4452...L10
Corbin, 7304...M14
Corydon, 720...H3
Covington, 40640...C12
Crab Orchard, 841...K13
Crescent Sprs., 3801...Y2*
Crestview, 475...Y4*
Crestview Hills, 3148...Y2*
Crestwood, 4531...F9
Crittenden, 3815...D12
Crofton, 749...L3
Cumberland, 2237...L17
Cynthiana, 6402...F13
Danville, 16218...J12
Dawson Sprs., 2764...L3
Dayton, 5338...X3*
Dixon, 786...J3
Douglass Hills, 5484...B10
Drakesboro, 515...L5
Dry Ridge, 2191...E12
Earlington, 1413...K3
E. Bernstadt, 716...L14
Eddyville, 2554...L1
Edgewood, 8575...Y2*
Edmonton, 1595...M9
Elizabethtown, 28531...J8
Elkhorn City, 982...K19
Elkton, 2062...M4
Elsmere, 8451...D12
Eminence, 2498...F10
Erlanger, 18082...C12
Evarts, 962...M17
Falmouth, 2169...E13
Fancy Farm, 458...F3
Ferguson, 924...L12
Fincastle, 817...A10
Flat Lick, 960...M15
Flatwoods, 7423...E18
Fleming-Neon, 770...K18
Flemingsburg, 2658...F15
Florence, 29951...D12
Fordsville, 524...J6
Forest Hills, 444...B9
Ft. Mitchell, 8207...C12
Ft. Thomas, 16325...X3*
Ft. Wright, 5723...Y2*
Fox Chase, 447...G9
Frankfort, 25527...G11
Franklin, 8408...M6
Fredonia, 401...L1
Frenchburg, 486...H15
Fulton, 2445...G3
Gamaliel, 376...N8
Garrison, 866...E17
Georgetown, 29098...G12
Ghent, 323...E11
Glasgow, 14028...L8
Glenview, 531...A8
Goshen, 909...F9
Graymoor-Devondale, 2870...A9
Grayson, 4217...F17
Greensburg, 2163...K9
Greenup, 1188...E18
Greenville, 4312...L4
Guthrie, 1419...N4
Hanson, 742...K3
Hardin, 615...M1
Hardinsburg, 2343...H6
Harlan, 1745...M16
Harrodsburg, 8340...H12
Hartford, 2672...K5
Hawesville, 945...H5
Hazard, 4456...K17
Hazel, 410...G5
Hebron, 5929...C12
Hebron Estates, 1087...G9
Henderson, 28757...H3
Hendron, 4687...E4
Hickman, 2395...F2
Highland Hts., 6923...Y3*
Hillview, 8172...G9
Hindman, 777...K17
Hodgenville, 3206...J9
Hollow Creek, 783...D9
Hollyvilla, 537...G9
Hopkinsville, 31577...M3
Horse Cave, 2311...L8
Houston Acres, 507...B9
Hurstbourne, 4216...B9
Hurstbourne Acres, 1811...B9
Hyden, 365...L16
Independence, 24757...D12
Indian Hills, 2868...A8
Inez, 717...H19
Irvine, 2715...J14
Irvington, 1181...H7
Island, 458...K4
Jackson, 2231...J16
Jamestown, 1794...L11
Jeffersontown, 26595...G9
Jeffersonville, 1506...G15
Jenkins, 2203...L18
Junction City, 2241...J12
Kevil, 376...D3
Kuttawa, 649...L1
La Center, 1009...D3
La Grange, 8082...F10
Lakeside Park, 2668...Y2*
Lancaster, 3442...J12
Langdon Place, 936...A10
Lawrenceburg, 10505...G11
Lebanon, 5539...J10
Lebanon Jct., 1813...H9
Ledbetter, 1683...E4
Leitchfield, 6699...K7
Lewisburg, 810...L5
Lewisport, 1670...H5
Lexington, 295803...G13
Liberty, 2168...K11
Livermore, 1365...J4
London, 7993...L14
Loretto, 713...J10
Louisa, 2467...G18
Louisville, 597337...G8
Loyall, 1461...M16
Ludlow, 4407...X2*
Lynch, 747...
Lyndon, 11002...
Lynnview, 914...
Madisonville, 19591...
Magnolia, 524...
Manchester, 1255...
Marion, 3039...
Martin, 634...
Masonville, 1014...
Mayfield, 10024...
Maysville, 9011...
McHenry, 388...
McKee, 800...
McRoberts, 784...
Meadow Vale, 736...
Melbourne, 401...
Middlesboro, 10334...
Middletown, 7218...
Midway, 1641...
Millersburg, 792...
Millstone, 117...
Milton, 574...E
Monticello, 6188...M
Moorland, 431...A
Morehead, 6845...G
Morganfield, 3285...
Morgantown, 2394...
Mortons Gap, 863...
Mt. Olivet, 299...E
Mt. Sterling, 6895...G
Mt. Vernon, 2477...K
Mt. Washington, 9117...
Muldraugh, 947...
Munfordville, 1615...
Murray, 17741...
Murray Hill, 582...
New Castle, 912...F
New Haven, 855...
Newport, 15273...C
Nicholasville, 28015...H
N. Corbin, 1773...L
Northfield, 1020...
N. Middletown, 643...G
Nortonville, 1204...
Oak Grove, 7489...
Olive Hill, 1599...F
Owensboro, 57265...
Owenton, 1327...E
Owingsville, 1530...G
Paducah, 25024...
Paintsville, 3459...H
Paris, 8553...G
Park City, 537...
Park Hills, 2970...Y
Parkway Vil., 650...
Pembroke, 869...
Perryville, 751...J
Petersburg, 620...C
Pewee Valley, 1456...
Phelps, 893...J
Pikeville, 6903...J
Pine Knot, 1621...N
Pineville, 1732...M
Pioneer Vil., 2030...
Plantation, 832...
Pleasureville, 834...F
Plum Sprs., 453...
Powderly, 745...
Prestonsburg, 3255...J
Princeton, 6329...
Prospect, 4698...
Providence, 3193...
Raceland, 2424...E
Radcliff, 21688...
Ravenna, 605...J
Reidland, 4491...
Richmond, 31364...H
River Bluff, 403...
Riverwood, 446...
Robards, 515...
Rolling Fields, 646...
Rolling Hills, 959...A
Russell, 3380...E
Russell Sprs., 2441...L
Russellville, 6960...M
Ryland Hts., 1022...D
Sacramento, 468...
St. Matthews, 17472...
St. Regis Park, 1454...
Salem, 752...
Salyersville, 1883...H
Sandy Hook, 675...G
Science Hill, 693...L
Scottsville, 4226...M
Sebree, 1603...
Seneca Gardens, 696...
Shelbyville, 14045...G
Shepherdsville, 11222...
Shively, 15264...
Silver Grove, 1102...C
Simpsonville, 2484...G
Smithland, 301...
Smiths Grove, 714...
Somerset, 11196...L
Southgate, 3803...X
S. Shore, 1122...E
S. Wallins, 859...M
S. Williamson, 602...J
Spottsville, 325...
Springfield, 2519...J
Spring Mill, 287...
Spring Valley, 654...
Stamping Ground, 643...F
Stanford, 3487...J
Stanton, 2733...H
Stearns, 1416...N
Strathmoor Vil., 648...
Sturgis, 1898...
Symsonia, 615...
Taylor Mill, 6604...Y
Taylorsville, 763...G
Tompkinsville, 2402...N
Trenton, 384...M
Union, 5379...D
Uniontown, 1002...
Upton, 683...
Vanceburg, 1518...E
Versailles, 8568...G
Villa Hills, 7489...C
Vine Grove, 4520...
Walton, 3635...D
Warsaw, 1615...D
Watterson Park, 976...
Wellington, 565...
W. Buechel, 1230...
W. Liberty, 3435...H
W. Point, 797...
Westwood, 4746...E
Wheelwright, 780...K
White Plains, 884...
Whitesburg, 2139...L
Whitesville, 552...
Whitley City, 1170...M
Wickliffe, 688...
Wilder, 3035...Y
Williamsburg, 5245...M
Williamstown, 3925...E
Wilmore, 3686...H
Winchester, 18368...G
Windy Hills, 2385...
Wingo, 632...
Woodlawn Park, 942...
Worthington, 1609...E
Wurtland, 995...E

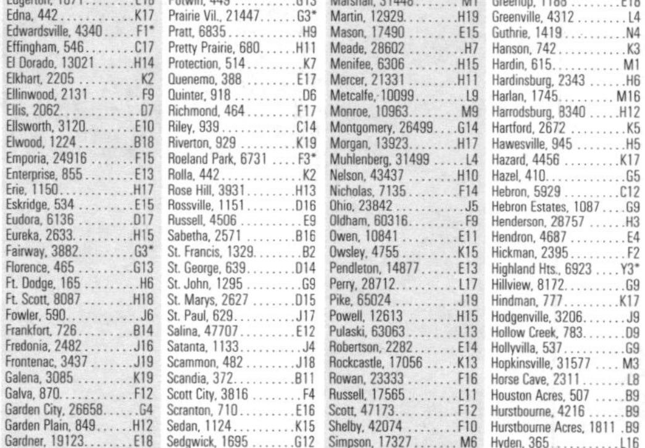

LOUISIANA

PG. 46–47

CAPITAL
Baton Rouge

NICKNAME
Pelican State

POPULATION
4,657,757, rank 25

AREA
47,751 sq mi, rank 3

STATEHOOD
1812, rank 18

Parishes

Acadia, 61773...
Allen, 25764...
Ascension, 107215...J
Assumption, 23421...L
Avoyelles, 42073...
Beauregard, 35654...
Bienville, 14353...
Bossier, 116979...
Caddo, 254969...
Calcasieu, 192768...
Caldwell, 10132...
Cameron, 6839...
Catahoula, 10407...E
Claiborne, 17195...
Concordia, 20822...F
De Soto, 26656...
E. Baton Rouge, 440171...H

MAINE
PG. 48–49

CAPITAL
Augusta

NICKNAME
Pine Tree State

POPULATION
1,362,359, rank 42

AREA
33,265 sq mi, rank 39

STATEHOOD
1820, rank 23

Counties

Cities and Towns

MARYLAND
PG. 50–51

CAPITAL
Annapolis

NICKNAME
Old Line State

POPULATION
6,177,224, rank 18

AREA
10,460 sq mi, rank 42

STATEHOOD
1788, rank 7

Counties

Cities and Towns
* City indexed to pg. 52
# Independent city population not included in county figures.

MASSACHUSETTS
PG. 53–55

CAPITAL
Boston

NICKNAME
Bay State

POPULATION
7,029,917, rank 15

AREA
8,284 sq mi, rank 45

STATEHOOD
1788, rank 6

Counties

Cities and Towns
* City indexed to pg. 52

MICHIGAN
PG. 56–58

CAPITAL
Lansing

NICKNAME
Great Lakes State

POPULATION
10,077,331, rank 10

AREA
58,527 sq mi, rank 23

STATEHOOD
1837, rank 26

Counties

Cities and Towns

MINNESOTA
PG. 59–61

CAPITAL
St. Paul

NICKNAME
Gopher State

POPULATION
5,706,494, rank 22

AREA
84,402 sq mi, rank 12

STATEHOOD
1858, rank 32

Counties

Cities and Towns
* City indexed to pg. 59

MISSISSIPPI
PG. 62

CAPITAL
Jackson

NICKNAME
Magnolia State

POPULATION
2,961,279, rank 34

AREA
47,689 sq mi, rank 32

STATEHOOD
1817, rank 20

Counties

MINNESOTA STATE CAPITOL, ST. PAUL, MINNESOTA

PEMAQUID POINT LIGHTHOUSE, MAINE

MONTANA PG. 66–67
CAPITAL
Helena
NICKNAME
Treasure State
POPULATION
1,084,225, rank 44
AREA
147,046 sq mi, rank 4
STATEHOOD
1889, rank 41

MISSOURI PG. 63–65
CAPITAL
Jefferson City
NICKNAME
Show Me State
POPULATION
6,154,913, rank 19
AREA
69,697 sq mi, rank 19
STATEHOOD
1821, rank 24

GLACIER NATIONAL PARK, MONTANA

NEBRASKA PG. 68–69
CAPITAL
Lincoln
NICKNAME
Cornhusker State
POPULATION
1,961,504, rank 37
AREA
77,355 sq mi, rank 15
STATEHOOD
1867, rank 37

NEVADA PG. 70
CAPITAL
Carson City
NICKNAME
Silver State
POPULATION
3,104,614, rank 32
AREA
110,561 sq mi, rank 7
STATEHOOD
1864, rank 36

COVERED WAGONS AT SCOTTS BLUFF NATL. MON., NEBRASKA

NEW HAMPSHIRE PG. 71
CAPITAL
Concord
NICKNAME
Granite State
POPULATION
1,377,529, rank 41
AREA
9,279 sq mi, rank 44
STATEHOOD
1788, rank 9

NEW JERSEY PG. 72–73
CAPITAL
Trenton
NICKNAME
Garden State
POPULATION
9,288,994, rank 11
AREA
7,787 sq mi, rank 46
STATEHOOD
1787, rank 3

NEW MEXICO PG. 74–75
CAPITAL
Santa Fe
NICKNAME
Land of Enchantment
POPULATION
2,117,522, rank 36
AREA
121,593 sq mi, rank 5
STATEHOOD
1912, rank 47

NEW YORK PG. 76–81
CAPITAL
Albany
NICKNAME
Empire State
POPULATION
20,201,249, rank 4
AREA
49,108 sq mi, rank 30
STATEHOOD
1788, rank 11

MANHATTAN BRIDGE, NEW YORK CITY, NEW YORK

CAPE HATTERAS BEACH, OUTER BANKS, NORTH CAROLINA

PENNSYLVANIA
PG. 94–98
CAPITAL
Harrisburg
NICKNAME
Keystone State
POPULATION
13,002,700, rank 5
AREA
45,308 sq mi, rank 33
STATEHOOD
1787, rank 2

RHODE ISLAND
PG. 99
CAPITAL
Providence
NICKNAME
Ocean State
POPULATION
1,097,379, rank 43
AREA
1,212 sq mi, rank 50
STATEHOOD
1790, rank 13

SOUTH CAROLINA
PG. 100
CAPITAL
Columbia
NICKNAME
Palmetto State
POPULATION
5,118,425, rank 23
AREA
31,113 sq mi, rank 40
STATEHOOD
1788, rank 8

SOUTH DAKOTA
PG. 101
CAPITAL
Pierre
NICKNAME
Mount Rushmore State
POPULATION
886,667, rank 46
AREA
77,116 sq mi, rank 16
STATEHOOD
1889, rank 40

TENNESSEE
PG. 102–103
CAPITAL
Nashville
NICKNAME
Volunteer State
POPULATION
6,910,840, rank 16
AREA
42,144 sq mi, rank 34
STATEHOOD
1796, rank 16

TEXAS
PG. 104–108
CAPITAL
Austin
NICKNAME
Lone Star State
POPULATION
29,145,505, rank 2
AREA
266,807 sq mi, rank 2
STATEHOOD
1845, rank 28

NEWPORT BRIDGE, NEWPORT, RHODE ISLAND

OLD BARN, OREGON

DELICATE ARCH, ARCHES NATIONAL PARK, UTAH

UTAH
PG. 109

CAPITAL
Salt Lake City

NICKNAME
Beehive State

POPULATION
3,271,616, rank 30

AREA
84,899 sq mi, rank 11

STATEHOOD
1896, rank 45

Counties

Cities and Towns

VIRGINIA
PG. 111–113

CAPITAL
Richmond

NICKNAME
Old Dominion

POPULATION
8,631,393, rank 12

AREA
40,767 sq mi, rank 36

STATEHOOD
1788, rank 10

Counties

Cities and Towns

VERMONT
PG. 110

CAPITAL
Montpelier

NICKNAME
Green Mountain State

POPULATION
643,077, rank 49

AREA
9,614 sq mi, rank 43

STATEHOOD
1791, rank 14

Counties

Cities and Towns

WASHINGTON
PG. 114–116

CAPITAL
Olympia

NICKNAME
Evergreen State

POPULATION
7,705,281, rank 13

AREA
68,138 sq mi, rank 20

STATEHOOD
1889, rank 42

Counties

Cities and Towns
* City indexed to pg. 111
# Independent city population in county figures

MOUNT RAINIER NATIONAL PARK, WASHINGTON

WEST VIRGINIA
PG. 117

CAPITAL
Charleston

NICKNAME
Mountain State

POPULATION
1,793,716, rank 39

AREA
24,231 sq mi, rank 41

STATEHOOD
1863, rank 35

Counties

Cities and Towns
* City indexed to pg. 50

WISCONSIN
PG. 118–119

CAPITAL
Madison

NICKNAME
Badger State

POPULATION
5,893,718, rank 20

AREA
56,153 sq mi, rank 26

STATEHOOD
1848, rank 30

Counties

Cities and Towns

WYOMING
PG. 120

CAPITAL
Cheyenne

NICKNAME
Equality State

POPULATION
576,851, rank 50

AREA
97,809 sq mi, rank 9

STATEHOOD
1890, rank 44

Counties
Albany, 36299 ....F11
Bear River, 518 ....A8

Canada

ALBERTA
PG. 123

CAPITAL
Edmonton

POPULATION
4,067,175, rank 4

AREA
255,541 sq mi, rank 6

ENTERED CANADA
1905

BRITISH COLUMBIA
PG. 122–123

CAPITAL
Victoria

POPULATION
4,648,055, rank 3

AREA
364,764 sq mi, rank 5

ENTERED CANADA
1871

MANITOBA
PG. 125

CAPITAL
Winnipeg

POPULATION
1,278,365, rank 5

AREA
250,116 sq mi, rank 8

ENTERED CANADA
1870

NEW BRUNSWICK
PG. 130

CAPITAL
Fredericton

POPULATION
747,101, rank 8

AREA
28,150 sq mi, rank 11

ENTERED CANADA
1867

NEWFOUNDLAND AND LABRADOR
PG. 131

CAPITAL
St. John's

POPULATION
519,716, rank 9

AREA
156,453 sq mi, rank 10

ENTERED CANADA
1949

NORTHWEST TERRITORIES
PG. 121

CAPITAL
Yellowknife

POPULATION
41,786, rank 11

AREA
519,734 sq mi (est.), rank 3

ESTABLISHED
1870

NUNAVUT
PG. 121

CAPITAL
Iqaluit

POPULATION
35,944, rank 12

AREA
808,185 sq mi, rank 1

ESTABLISHED
1999

ENTERED CANADA
1867

NOVA SCOTIA
PG. 130–131

CAPITAL
Halifax

POPULATION
923,598, rank 7

AREA
21,345 sq mi, rank 12

ENTERED CANADA
1867

ONTARIO
PG. 125–127

CAPITAL
Toronto

POPULATION
13,448,494, rank 1

AREA
415,598 sq mi, rank 4

ENTERED CANADA
1867

PRINCE EDWARD ISLAND
PG. 130–131

CAPITAL
Charlottetown

POPULATION
142,907, rank 10

AREA
2,185 sq mi, rank 13

ENTERED CANADA
1873

QUEBEC
PG. 128–130

CAPITAL
Québec

POPULATION
8,164,361, rank 2

AREA
595,391 sq mi, rank 2

ENTERED CANADA
1867

SASKATCHEWAN
PG. 124–125

CAPITAL
Regina

POPULATION
1,098,352, rank 6

AREA
251,366 sq mi, rank 7

ENTERED CANADA
1905

YUKON
PG. 121

CAPITAL
Whitehorse

POPULATION
35,874, rank 13

AREA
186,272 sq mi, rank 5

ESTABLISHED
1898

Mexico

MEXICO
PG. 132–133

CAPITAL
Mexico City

POPULATION
112,336,538

AREA
756,066 sq mi

Puerto Rico

PUERTO RICO
PG. 134

CAPITAL
San Juan

POPULATION
3,285,874

AREA
3,435 sq mi

OURISM INFORMATION

NITED STATES

bama
w.alabama.travel
veetHomeAla

ska
w.travelalaska.com
ravelAlaska

rizona
ww.visitarizona.com
ArizonaTourism

rkansas
www.arkansas.com
@artourism

California
www.visitcalifornia.com
@VisitCA

Colorado
www.colorado.com
@Colorado

Connecticut
www.ctvisit.com
@CTvisit

laware
w.visitdelaware.com
elawareTourism

strict of Columbia
vashington.org
washingtondc

lorida
www.visitflorida.com
VisitFlorida

gia
exploregeorgia.org
reGeorgia

'i
v.gohawaii.com
gohawaii

daho
visitidaho.org
@visitidaho

Illinois
www.enjoyillinois.com
@enjoyillinois

Indiana
visitindiana.com
@visitindiana

Iowa
www.traveliowa.com
@Travel_Iowa

Kansas
www.travelks.com
@TravelKS

Kentucky
www.kentuckytourism.com
@KentuckyTourism

Louisiana
www.louisianatravel.com
@LouisianaTravel

Maine
visitmaine.com
@visitmaine

Maryland
www.visitmaryland.org
@TravelMD

Massachusetts
visitma.com
@VisitMA

Michigan
www.michigan.org
@PureMichigan

Minnesota
www.exploreminnesota.com
@ExploreMinn

Mississippi
www.visitmississippi.org
@visitms

Missouri
www.visitmo.com
@VisitMO

Montana
www.visitmt.com
@visitmontana

Nebraska
www.visitnebraska.com
@NebraskaTourism

Nevada
www.travelnevada.com
@TravelNevada

New Hampshire
www.visitnh.gov
@VisitNH

New Jersey
www.visitnj.org
@Visit_NJ

New Mexico
www.newmexico.org
@NewMexico

New York
www.iloveny.com
@I_LOVE_NY

North Carolina
www.visitnc.com
@VisitNC

North Dakota
www.ndtourism.com
@NorthDakota

Ohio
www.ohio.org
@OhioFindItHere

Oklahoma
www.travelok.com
@TravelOK

Oregon
traveloregon.com
@TravelOregon

Pennsylvania
www.visitpa.com
@visitPA

Puerto Rico
www.discoverpuertorico.com
@discover_PR

Rhode Island
www.visitrhodeisland.com
@RITourism

South Carolina
www.discoversouthcarolina.com
@Discover_SC

South Dakota
www.travelsouthdakota.com
@southdakota

Tennessee
www.tnvacation.com
@TNVacation

Texas
www.traveltexas.com
@TravelTexas

Utah
www.visitutah.com
@VisitUtah

Vermont
www.vermontvacation.com
@VermontTourism

Virginia
www.virginia.org
@VisitVirginia

Virgin Islands
www.visitusvi.com
@USVItourism

Washington
www.experiencewa.com
@ExperienceWA

West Virginia
wvtourism.com
@WVtourism

Wisconsin
www.travelwisconsin.com
@TravelWI

Wyoming
www.travelwyoming.com
@visitwyoming

CANADA

Alberta
www.travelalberta.com
@TravelAlberta

British Columbia
www.hellobc.com
@HelloBC

Manitoba
www.travelmanitoba.com
@TravelManitoba

New Brunswick
www.tourismnewbrunswick.ca
@DestinationNB

Newfoundland & Labrador
www.newfoundlandlabrador.com
@NLtweets

Northwest Territories
www.spectacularnwt.com
@spectacularNWT

Nova Scotia
www.novascotia.com
@VisitNovaScotia

Nunavut
www.travelnunavut.ca
@TravelNunavut

Ontario
www.ontariotravel.net
@OntarioTravel

Prince Edward Island
www.tourismpei.com
@tourismpei

Québec
www.bonjourquebec.com
@TourismQuebec

Saskatchewan
www.tourismsaskatchewan.com
@Saskatchewan

Yukon
www.travelyukon.com
@TravelYukon

MEXICO

Mexico
www.visitmexico.com
@VisitMex

ROAD CONSTRUCTION & CONDITIONS

UNITED STATES

In many parts of the U.S., travelers may now **dial 511** for current information on state and local road construction and road conditions.

For more details, visit the website at *www.fhwa.dot.gov/trafficinfo/511.htm*

Consult the following phone numbers and websites for information about road construction and road conditions.

Alabama
algotraffic.com

Alaska
511 or 866.282.7577
511.alaska.gov

Arizona
511 or 888.411.7623
www.az511.com

Arkansas
501.569.2000
www.idrivearkansas.com

California
511 or 800.427.7623
quickmap.dot.ca.gov

Colorado
511 or 303.639.1111
www.cotrip.org

Connecticut
860.594.2000
cttravelsmart.org

Delaware
800.652.5600
www.deldot.gov

District of Columbia
202.673.6813
ddot.dc.gov

Florida
511
fl511.com

Georgia
511 or 877.694.2511
www.511ga.org

Hawai'i
www.hidot.hawaii.gov

Idaho
511 or 888.432.7623
511.idaho.gov

Illinois
800.452.4368
www.gettingaroundillinois.com

Indiana
855.463.6848
pws.trafficwise.org

Iowa
511 or 800.288.1047
www.511ia.org

Kansas
511 or 866.511.5368
www.kandrive.org

Kentucky
511 or 877.367.5982
goky.ky.gov

Louisiana
511 or 888.762.3511
www.511la.org

Maine
511 or 207.624.3000
newengland511.org

Maryland
511 or 855.466.3511
chart.maryland.gov

Massachusetts
511
mass511.com

Michigan
517.335.3084
mdotjboss.state.mi.us/MiDrive/map

Minnesota
511 or 800.542.0220
511mn.org

Mississippi
511 or 866.521.6368
www.mdottraffic.com

Missouri
888.275.6636
traveler.modot.org/map

Montana
511 or 800.226.7623
roadreport.mdt.mt.gov

Nebraska
511 or 800.906.9069
www.511.nebraska.gov

Nevada
511 or 402.471.4567
www.nvroads.com

New Hampshire
511 or 603.271.6862
newengland511.org

New Jersey
511 or 866.511.6538
www.511nj.org

New Mexico
511 or 800.432.4269
nmroads.com

New York
511 or 888.465.1169
www.511ny.org

North Carolina
511 or 877.511.4662
drivenc.org

North Dakota
511 or 866.696.3511
www.travel.dot.nd.gov

Ohio
614.466.7170
www.ohgo.com

Oklahoma
405.522.8000
ok.gov/odot

Oregon
511 or 800.977.6368
www.tripcheck.com

Pennsylvania
511 or 877.511.7366
www.511pa.com

Rhode Island
511 or 844.368.7623
www.dot.ri.gov/travel

South Carolina
511 or 855.467.2368
www.511sc.org

South Dakota
511 or 605.773.3265
www.sd511.org

Tennessee
511 or 877.244.0065
smartway.tn.gov

Texas
800.452.9292
drivetexas.org

Utah
511 or 866.511.8824
udottraffic.utah.gov

Vermont
511 or 802.476.2690
newengland511.org

Virginia
511 or 800.367.7623
www.511virginia.org

Washington
511 or 800.695.7623
www.wsdot.com/traffic

West Virginia
511 or 855.699.8511
www.wv511.org

Wisconsin
511 or 866.511.9472
www.511wi.gov

Wyoming
511 or 888.996.7623
map.wyoroad.info

CANADA

Alberta
511 or 855.391.9743
511.alberta.ca

British Columbia
800.550.4997
www.drivebc.ca

Manitoba
511 or 866.626.4862
www.manitoba511.ca

New Brunswick
511 or 800.561.4063
511.gnb.ca

Newfoundland & Labrador
709.729.2300
www.511nl.ca

Northwest Territories
800.661.0750
www.dot.gov.nt.ca

Nova Scotia
511 or 888.780.4440
511.novascotia.ca/en

Ontario
511 or 866.929.4257
511.on.ca

Prince Edward Island
511 or 855.241.2680
511.gov.pe.ca/en

Québec
511 or 888.355.0511
www.quebec511.info

Saskatchewan
888.335.7623
hotline.gov.sk.ca

Yukon
511 or 867.667.5811
www.511yukon.ca

MEXICO

www.gob.mx/sct
(limited English content)

BORDER CROSSING

TRAVEL ADVISORY

All U.S. citizens are now required to present a passport, passport card, or WHTI (Western Hemisphere Travel Initiative)-compliant document when entering the United States by air, sea or land. U.S. citizens traveling directly to or from Puerto Rico and the U.S. Virgin Islands are not required to have a passport. For more detailed information and updated schedules, please see http://travel.state.gov.

CANADA

Canadian law requires that all persons entering Canada carry both proof of citizenship and proof of identity. A valid U.S. passport, passport card or other WHTI-compliant document satisfies these requirements for U.S. citizens. U.S. citizens entering Canada from a third country must have a valid U.S. passport. A visa is not required for U.S. citizens to visit Canada for up to 180 days.

U.S. driver's licenses are valid in Canada. Drivers should be prepared to present proof of their vehicle's registration, ownership, and insurance.

UNITED STATES (FROM CANADA)

Canadian citizens are required to present valid WHTI-compliant documents when entering the United States by land or water. These documents include a passport or an Enhanced Driver's Licence/Enhanced Identification Card. Canadian citizens traveling by air to, through or from the United States must present a valid passport (see Travel Advisory). Visas are not required for customary tourist travel.

Canadian driver's licenses are valid in the U.S. for one year. Drivers should be prepared to present proof of their vehicle's registration, ownership, and insurance.

MEXICO

All persons entering Mexico need either a valid passport or their original birth certificate along with a valid photo ID such as a drivers license (U.S. citizens should bear in mind the requirements set by the U.S. government for re-entry to the U.S.). Visas are not required for stays of up to 180 days. Naturalized citizens and alien permanent residents should carry the appropriate official documentation. Individuals under the age of 18 traveling alone, with one parent, or with other adults must carry notarized parental/legal guardian authorization. All U.S. citizens visiting for up to 180 days must also procure a tourist card, obtainable from Mexican consulates, tourism offices, border crossing points, and airlines serving Mexico. However, tourist cards are not needed for visits shorter than 72 hours to areas within the Border Zone (extending approximately 25 km into Mexico)

U.S. driver's licenses are valid in Mexico. Visitors who wish to drive beyond the Baja California Peninsula or the Border Zone must obtain a temporary import permit for their vehicles. To acquire a permit, one must submit evidence of citizenship and of the vehicle's title and registration, as well as a valid driver's license. A processing fee must be paid. Permits are available at any Mexican Army Bank (Banjercito) located at border crossings or selected Mexican consulates. Mexican law also requires the posting of a refundable bond, via credit card or cash, at the Banjercito to guarantee the departure of the vehicle. Do not deal with any individual operating outside of official channels.

All visitors driving in Mexico should be aware that U.S. auto insurance policies are not valid and that buying short-term tourist insurance is mandatory. Many U.S. insurance companies sell Mexican auto insurance. American Automobile Association (for members only) and Sanborn's Mexico Insurance (800.638.9423) are popular companies with offices at most U.S. border crossings.

Published by
National Geographic Maps
in association with Globe Turner, LLC.

This product contains proprietary property of Globe Turner, LLC and National Geographic Maps. Reproduction or recording of any maps, photographs, tables, text, or other material contained in this publication in any manner including, without limitation, by photocopying and electronic storage and retrieval, is prohibited.

No part of this National Geographic Road Atlas may be reproduced in any form without the prior written consent of both Globe Turner, LLC and National Geographic Maps.

The information contained in this product is derived from a variety of third party sources. While every effort has been made to verify the information contained in such sources, National Geographic Maps and Globe Turner, LLC assume no responsibility for inconsistencies or inaccuracies in the data, nor liability for any damages of any type arising from errors or omissions. The Adventure section serves only as a guide to activities. Please check with local authorities for conditions and safety issues.

Copyright © 2022
by Globe Turner, LLC.
All rights reserved.
Exclusive license held by
Mapping Specialists, Ltd.

Adventure section
Copyright © 2022
National Geographic Partners, LLC
All rights reserved.

NATIONAL GEOGRAPHIC and
Yellow Border Design
are trademarks of the
National Geographic Society,
used under license.

Printed in Canada
ISBN 978-0-79228-989-0

Comments or suggestions for the Adventure section should be directed to:
National Geographic Maps
212 Beaver Brook Canyon Road
Evergreen, CO 80439
1.800.962.1643

Adventure text:
Cliff Ransom,
National Geographic Adventure

Adventure photography:
Pg. C: Colorado River, AZ, Tom Bean; **Pg. D-E:** Adirondack Park, NY, Rob Howard; Delaware Water Gap NRA, PA, Skip Brown; Youghiogheny River Gorge, PA, Skip Brown; Hawk Mountain Sanctuary, PA, A.E. Morris/Birds as Art; Asheville, NC, Alex Di Suvero; **Pg. F-G:** Wichita Mountains NWR, OK, Harrison Shull; Big Bend NP, TX, Tom Bean/Corbis; Cable, WI, Lynn Howell/American Birkebeiner; Mississippi River Trail, IA, Barry Tessman; Texas Gulf Coast, Doug Wechsler/VIREO; **Pg. H-I:** Grand Canyon NP, AZ , Russell Kaye; Verde River, AZ, Mike Padian; Ouray, CO, Daniel H. Bailey; Moab, UT, Steve Casimiro; Bosque del Apache NWR, NM, Cliff Beittel; **Pg. J-K:** Death Valley NP, CA, Cliff Leight; San Juan Islands, WA, Roy Toft; Chugach Mountains, AK, Mark Gamba; Lake Tahoe, CA, Rich Reid/ Colors of Nature.

National Parks photography:
Arkansas Dept. of Tourism (Hot Springs); Craig Haggit (Banff, Grand Canyon); National Park Service (Acadia, Arches, Mammoth Cave, Rocky Mountain, Shenandoah, Voyageurs); Getty Images, Inc. (all others)

Index photography:
© Corbis and Getty Images, Inc.

Interstate Route
Other Route

206   Distance in Miles
4:15   Approximate Travel Time
● Miami   City on Mileage Chart (inside back cover)
● Fort Pierce   Other City

Distances and driving times may vary depending on actual route traveled and driving conditions.

©Globe Turner